MODERN GENERAL CHEMISTRY LABORATORY

INCORPORATING COMPUTER-ORIENTED DATA ACQUISITION AND EVALUATION APPROACH INTO THE STUDENT LABORATORY EXPERIENCE

WILLIAM E. ACREE, Jr.

i

Printed in the United States of America

ISBN13: 978-9838310-3-7

UNIVERSITY OF NORTH TEXAS
Eagle Images
1155 Union Circle
#309615
Denton, TX 76203

940-565-2083

Address all correspondence and order information to the above address.

Cover photos: roy nance

PREFACE

The laboratory manual is designed for a two-semester general chemistry laboratory sequence for science majors. The laboratory experiments that have been included cover most (if not all) of the topics typically found in university-level general chemistry laboratory courses throughout the United States. Topics covered include the properties of matter, stoichiometric relationships, chemical formula, molar mass determination, chemical syntheses, properties of gases, various statistical analyses, thermochemistry, phase behavior, colligative properties, acid-base and oxidation-reduction chemical reactions, chemical kinetics, chemical equilibrium, spectroscopic measurements and electrochemistry. Approximately half of the experiments use the Pasco probeware and data acquisition system (Version 1.9.0). Using probeware, students are able to make temperature, absolute pressure, pH, potential and absorbance measurements. Measured values are displayed in "real time", and using built-in software functions, students can perform various mathematical computations and statistical analyses on their recorded experimental data. Although the specific instructions were written with the Pasco system in mind, the experiments can easily be adapted to other probeware and data acquisition systems, such as the Vernier system and MeasureNet system.

TABLE OF CONTENTS

CHEMICAL LABORATORY SAFETY RULES

Depending on your major, you will be enrolled in the general chemistry laboratory course for either one semester or for two semesters. Each semester you will perform twelve laboratory experiments, each of which is designed to illustrate different chemical principles and concepts, to apply several computational methods, and to learn various experiment techniques. You will be exposed to many different chemical substances, and will learn how to use various pieces of chemical glassware and chemical instrumentation. It is essential that you read the laboratory experiments prior to going to the laboratory, and that you abide by all of the chemical laboratory safety rules that are typed below. These rules are designed to ensure that work performed in the laboratory will be safe for you and your fellow students. After you have carefully read and understood the safety rules, please sign the attached form and give it to your Teaching Assistant. There should be two copies of the form, one copy to signed and turned in this semester, and the second copy is for next semester's course.

SAFETY RULES

1. Never enter the laboratory unless your Teaching Assistant (TA) is present. Accidents are more likely to occur when students are left unsupervised in the chemistry laboratory.
2. Always wear approved safety goggles/glasses when in the chemistry laboratory. Regular vision glasses are not the same as chemical safety glasses. Vision glasses do not afford eye protection from the side the way that chemical safety glasses do.
3. Always read the laboratory experiment ahead of time – and if you do have questions regarding safety or the experimental procedure, ask TA before attempting to perform the experiment.
4. Never eat, drink, chew or smoke in the laboratory.
5. Wear sensible, relatively protective clothing. Open-toed shoes, sandals, *etc.* are not allowed, nor are ultra short shorts, mini-skirts, *etc.* The TAs have been instructed to send anyone home who does have the proper footwear.
6. Long hair should be tied back, and ties/scarfs, *etc.* should be removed.
7. Always wash your hands thoroughly before leaving the laboratory.
8. Horseplay and unauthorized experiments are totally forbidden in the laboratory. The TAs have been instructed to send anyone home who does not abide by this rule.
9. Place coats, jackets, books and backpacks in the book/backpack shelves located in the laboratory. Cluttered aisles and lab benches are dangerous.
10. All chemicals, glassware and instrumentation should be treated with the utmost respect.
11. Never smell or touch chemicals unless specifically instructed to so. When instructed to smell chemicals, hold the container level with your nose, but

removed by several inches, and waft the vapors towards you by waving your hand over the top of the container. Never put the container directly under your nose.

12. Chemical waste and broken glassware should be disposed of in the correct fashion. The laboratories have special containers for collecting chemical waste and broken glassware. The chemical waste containers will be labeled with the type of waste that is supposed to be put in the container. Read the label on the waste container before putting any chemical into the container. For safety reasons, some chemicals have to be disposed of in separate containers. The TA will provide more instruction on how you are to dispose of the chemicals for each laboratory experiment.

13. All accidents/breakages/spills/injuries should be reported to the TA immediately.

14. Familiarize yourself with the location of all safety equipment – fire extinguishers, safety shower, eye-wash station, eye-wash bottles, and first aid kit.

15. Carefully read the label on bottles for the identity of a substance before using the chemical. Abide by any warnings that might be on the label.

16. If you are unsure of any chemical, ask the TA or check the Materials Safety Data Sheets (MSDS) that are available in the Chemistry Stockroom for any chemical.

17. Never pipette by mouth.

18. Always clean up after yourself and keep your work area neat and clean at all times. When you leave the laboratory your work area should be clean and all common glassware returned to its storage location. The TA has been instructed to deduct points from that day's laboratory report if you leave without cleaning your work area.

19. If you break a thermometer, do not touch the mercury. **TELL YOUR TA IMMEDIATELY** and cover any mercury with sulfur powder. The Department has special kits for cleaning up any mercury that might be spilled.

20. Cover any acid spills with sodium bicarbonate ($NaHCO_3$).

21. Be aware of all hazard warnings before entering the laboratory and follow them.

SAFETY RULES

1. Never enter the laboratory unless your Teaching Assistant (TA) is present. Accidents are more likely to occur when students are left unsupervised in the chemistry laboratory.

2. Always wear approved safety goggles/glasses when in the chemistry laboratory. Regular vision glasses are not the same as chemical safety glasses. Vision glasses do not afford eye protection from the side the way that chemical safety glasses do.

3. Always read the laboratory experiment ahead of time – and if you do have questions regarding safety or the experimental procedure, ask TA before attempting to perform the experiment.

4. Never eat, drink, chew or smoke in the laboratory.

5. Wear sensible, relatively protective clothing. Open-toed shoes, sandals, *etc.* are not allowed, nor are ultra short shorts, mini-skirts, *etc.* The TAs have been instructed to send anyone home who does have the proper footwear.

6. Long hair should be tied back, and ties/scarfs, *etc.* should be removed.

7. Always wash your hands thoroughly before leaving the laboratory.

8. Horseplay and unauthorized experiments are totally forbidden in the laboratory. The TAs have been instructed to send anyone home who does not abide by this rule.

9. Place coats, jackets, books and backpacks in the book/backpack shelves located in the laboratory. Cluttered aisles and lab benches are dangerous.

10. All chemicals, glassware and instrumentation should be treated with the utmost respect.

11. Never smell or touch chemicals unless specifically instructed to so. When instructed to smell chemicals, hold the container level with your nose, but removed by several inches, and waft the vapors towards you by waving your hand over the top of the container. Never put the container directly under your nose.

12. Chemical waste and broken glassware should be disposed of in the correct fashion. The laboratories have special containers for collecting chemical waste and broken glassware. The chemical waste containers will be labeled with the type of waste that is supposed to be put in the container. Read the label on the waste container before putting any chemical into the container. For safety reasons, some chemicals have to be disposed of in separate containers. The TA will provide more instruction on how you are to dispose of the chemicals for each laboratory experiment.

13. All accidents/breakages/spills/injuries should be reported to the TA immediately.

14. Familiarize yourself with the location of all safety equipment – fire extinguishers, safety shower, eye-wash station, eye-wash bottles, and first aid kit.

15. Carefully read the label on bottles for the identity of a substance before using the chemical. Abide by any warnings that might be on the label.

16. If you are unsure of any chemical, ask the TA or check the Materials Safety Data Sheets (MSDS) that are available in the Chemistry Stockroom for any chemical.

17. Never pipette by mouth.

-OVER-

18.   Always clean up after yourself and keep your work area neat and clean at all times. When you leave the laboratory your work area should be clean and all common glassware returned to its storage location. The TA has been instructed to deduct points from that day's laboratory report if you leave without cleaning your work area.

19.   If you break a thermometer, do not touch the mercury. **TELL YOUR TA IMMEDIATELY** and cover any mercury with sulfur powder. The Department has special kits for cleaning up any mercury that might be spilled.

20.   Cover any acid spills with sodium bicarbonate (NaHCO$_3$).

21.   Be aware of all hazard warnings before entering the laboratory and follow them.

Printed Student Name: _____   Date: _____

Signature: _____

Student ID Number: _____

EXPERIMENT 1A: STATISTICAL ANALYSIS ON DIFFERENT TYPES OF PENNIES

INTRODUCTION

Every experimental measurement involves some level of experimental uncertainty or error. For example, if one was asked to repeatedly measure the length of an 8-foot table with a 12-inch ruler, one would not get exactly the same numerical value every time. There would be some level of uncertainty associated with placement of the ruler every time that one picked up the ruler and replaced it on the tabletop as one moves from the left-hand side of the table to the right-hand side. One would have to pick the ruler up several times in order to measure the table's length. Even if one were to make a small pencil or chalk mark on the tabletop to indicate where each ruler length ended, it would be extremely difficult (if not impossible) to reproducibly place the ruler in exactly the same spot every time in a series of replicate measurements. Experimental uncertainties (or errors) give rise to different numerical values when a series of replicate measurements are made on the same sample or object. In other words, there will be a dispersion (or spread) of the measured values. The smaller the spread of values, the more precise will be the average (or mean) value of the experimental measurements. Precision is a measure of the reproducibility (or scatter) of replicate measurements. Precision has absolutely nothing to do with how close the measured values are to the so-called true value. Accuracy is the term used to indicate how close a measured value is to the true or accepted value. The average of several replicate measurements might be accurate without being very precise. Or conversely, a set of replicate measurements might be very reproducible (*e.g.*, very precise), yet the measured values may not be very close to the true or accepted value. The ideal situation would be that the series of replicate measurements were both accurate and precise.

Typically when one performs a series of replicate measurements, one reports not only the individual values, but also the mean value (also referred to as the average value) and standard deviation. The mean value, $\bar{x}$, is obtained by adding all of the individual measurements and then dividing by the total number of measurements.

$$\bar{x} = \Sigma\, x_i\, /\, n \qquad\qquad\qquad (1A.1)$$

where Σ represents the mathematical summation operation, and n is the number of replicate measurements performed. The standard deviation is calculated as

1

$$s = \sqrt{\frac{\sum (x_i - \bar{x})^2}{n - 1}} \qquad (1A.2)$$

Standard deviations are generally expressed with a "±" sign; *e.g.* "100.53 ± 0.32". The smaller the standard deviation, the less scattered the numerical values. When the standard deviation is small, the individual values are grouped closer about the mean value.

The calculated standard deviation does have statistical significance in that random experimental errors follow a Gaussian error distribution (*e.g.*, "bell-shaped distribution"). If a given measurement is performed a large number of times, one will obtain a range of numerical values that will be distributed according to a Gaussian error distribution, provided that one does not make systematic errors of measurement. For random errors, small errors are much more probable than large errors, and positive deviations are just as likely as negative ones. In the laboratory, we rarely repeat the same experiment enough times to truly obtain a Gaussian distribution. We can take our smaller set of measured values; however, calculate the mean and standard deviation for our smaller data set, and using statistics project what the Gaussian distribution would look like, had we performed the much larger number of experimental measurements. The standard deviation is used to describe the variation in a finite data set, whereas the variance (denoted as σ), or population standard deviation as it is sometimes called, is used in an infinite population. The variance is calculated in a similar fashion as the standard deviation, except that the denominator is n, rather than n-1 (see Equation 1A.2).

The population standard deviation measures the width of the Gaussian distribution curve. The larger the value of σ, the broader is the curve. For any Gaussian distribution, 68.3 % of the area under the curve falls in the range of $\bar{x} \pm 1\sigma$. That is, more than two-thirds of the measurements are expected to lie within one standard deviation of the mean. Also, 95.5 % of the area under the Gaussian curve lies within $\bar{x} \pm 2\sigma$, and 99.7 % of the area lies within $\bar{x} \pm 3\sigma$. (See Figure 1A.1) Only a very, very small percentage of the experimental measurements would be expected to be more than 3 standard deviations from the mean as the result of random errors. One can use this rationalization in reverse. Suppose that one did measure an experimental value that was more than three standard deviations larger than the calculated average value for five replicate experimental determinations. The experiment had been performed five times, and one value was significantly different from the other values. What would you think or conclude based on the preceding discussion? That the difference between the one apparent "outlier" and the other four values resulted from solely random errors, or that one unknowingly made a mistake in that particular experimental measurement?

GAUSSIAN DISTRIBUTION

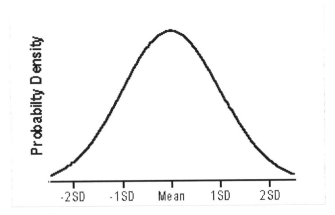

Figure 1A.1. Cumulative probability distribution of a Gaussian bell-shaped distribution. The standard probability distribution is scaled so that the area under the curve equals 1.0.

In Experiment 1 we will determine the average mass and standard deviation for a post-1982 collection of pennies, and using statistics determine whether or not there is a significant difference in the mass of a pre-1982 penny versus the mass of a post-1982 penny versus the mass of a 1943 steel penny. In 1982 the United States Mint changed the composition of the penny from a predominantly copper-based alloy (1963-1982 pennies are 95.0% Cu + 5.0% Zn; 1947-1962 pennies are 95.0% Cu + 5.0% (Zn & Sn)) to a zinc-based alloy (97.5% Zn + 2.5% Cu). The 1943 pennies were made of steel with zinc coating. The diameter of the penny has remained the same, 19 mm. If a difference is found, it will be caused by a change in the manufacturing process (*e.g.*, different alloy material), and not the uncertainty in measuring the mass of the penny. The error in the mass measurement should be quite small, if the electronic balance is functioning properly. The manufacturer's specification for our electronic balances is an accuracy of $\pm$ 2 in the last decimal place.

The type of statistical analysis that we will be performing is quite similar to what manufacturing companies use in assessing quality control, in doing hypothesis testing, and in verifying warranty claims. In manufacturing applications one would be applying statistical methods to determine whether or not a manufactured product met the given specification within the predetermined, set tolerance level.

STANDARD DEVIATION DETERMINATION ON POST-1982 PENNIES

Using an electronic balance determine the mass of 20 post-1982 pennies. Record the mass of each penny on your Laboratory Data Sheet for Experiment 1. Next calculate the average mass (mean) and standard deviation of the 20 post-1982 pennies. Most scientific hand-held calculators have a built-in statistic package that will perform this calculation. If you know how to use this function on your calculator, fine. Go ahead and calculate both numerical values and record the mean and standard deviation on the Data Sheet in the indicated space.

The Pasco system has the capability of performing statistical calculations. To calculate and average and standard deviation using the built-in software, double on Data Studio. You have now opened the system. **Double click on the activity titled Enter Data.** What should now appear on the screen is a table on the left-hand side, and a graph on the right-hand side. Enter the values to be averaged in the x-column of the table. Don't worry about what is happening with the y-column. Once all of the values are entered, **click on the Σ key** on the Table toolbar. The mean should now be displayed at the bottom of the column. If you want to calculate the standard deviation, go back up to the top of the table, and **click on the ▼ arrow by the Σ key**, scroll down to standard deviation, and hit enter. The standard deviation is now at the bottom the page, along with the average. When you get ready to enter the next set of data, you can either enter the next set right over the previous data set (be sure to delete out any extra data points), or you can clear the numerical values in the table by **mouse clicking on New Data** on the top tool bar. [Alternatively, you can bring down an empty data table by **mouse clicking on the Experiment key** on the top toolbar, and then **scrolling down to New Empty Data Table**. The latter method will probably take a few minutes longer. Since you have to type all of the numbers anyway, it is probably easier to type over existing numbers, than bringing down new tables and then typing the new numbers into an empty table.)

Microsoft Excel Instructions for Average and Standard Deviation Calculations: Some of the calculations during the semester will have to be performed outside of the laboratory time. This will be true of the average and standard deviation calculations based on the experimental data from the entire class. You can perform the calculation using Microsoft Excel spreadsheet. The general computer access laboratories across campus should have Microsoft Excel as part of the standard Microsoft Windows bundle of software. The Excel spreadsheet will appear as a grid of cells, which are identified by both a column alphabetical letter and row numerical value. In cells A5 through A24 enter the mass of the 20 individual post-1982 pennies. In cell A3 type: **=AVERAGE(A5:A24)** . Hit **ENTER** -By typing this command you are programming the computer to calculate the average value of the numbers that are entered in cells A5 through A24. The calculated mean (average value) should now appear in cell A3. To calculate the standard deviation of the numerical values stored in cells A5 through A24, go to cell B3 and type: **=STDEV(A5:A24).** Hit **ENTER** - You have just instructed the computer to calculate the standard deviation of the values that are entered in cells A5 through A24. The standard deviation should now appear in cell B3. Microsoft Excel has several built-in statistical features that we may need to access during the semester to complete parts of the laboratory experiments.

COMPARISON OF DIFFERENT DATA SETS USING STUDENT'S t-TEST

Using an electronic balance, weight five pre-1982 and five post-1982 pennies. Record the mass of the individual pennies on the laboratory data sheet. The masses of five 1943 steel pennies have already been recorded on the data sheet for you. The data sheets will guide you through the three mean and standard deviation calculations if your calculator does not have this built-in function. Alternatively, you can calculate the mean and standard deviation of each data set using Microsoft Excel spreadsheet. (The instructions for using Microsoft Excel are in the preceding section of this laboratory experiment. Enter the individual data points in cells A5 through A9 – delete what is in cells A10 through A24. You will also need to modify the mathematical formulas in cells A3 and B3. In cell A3 type - **=AVERAGE(A5:A9)** . Hit **ENTER** – The average value should now appear in cell A3. In cell B3 type - =STDEV(A5:A24) . Hit **ENTER** . The standard deviation for the data set in cells A5 through A9 should appear in cell B3.) Do this calculation for the three datasets.

Next we need to analyze the experimental results using statistics to determine how significant the difference is between the pre-1982 pennies versus post-1982 pennies, between the pre-1982 pennies versus 1943 steel pennies, and between post-1982 pennies versus 1943 steel pennies. You need to make three comparisons. Some idea of whether or not there is a significant difference between any two data sets can be gained simply by looking at the data sets being compared. Suppose that the two data sets had exactly the same average value to the third decimal place. What would your intuition tell you? That the two data sets were the same, at least as far as the mass is concerned. It is very unlikely though, that the two data sets will have exactly the same average value. Now what type of information would you want to have? You would probably want to know not only the average mass of each data set, but also how much variation was there in the individual masses for each data set. In other words, you would want to know the mean and standard deviation of both data sets being compared. As was stated earlier, we could use the calculated mean and standard deviation for each finite data set to construct a Gaussian distribution for the pre-1982 pennies, for the post-1982 pennies and for the 1943 steel pennies using statistic methods. If I were asked to compare the pre-1982 pennies versus the post-1982 pennies, I could look at the amount of overlap between the two respective Gaussian distributions. For example, if the two Gaussian distributions were super-imposable on one another, then I would likely state that the two sets of pennies are definitely identical, at least as far their mass is concerned. If there was a lot of overlap in the two Gaussian distributions, I would likely state that I believe that the two sets of pennies are identical. A little bit less overlap, that the two data sets may be identical. Even less overlap, that the two data sets are likely different. And finally to the point where there is no overlap between the two data sets, where I would likely conclude that the two data sets are indeed different. You will note that I was hedging my statement depending on how confident I was in the conclusion that I was making based on the analyses that I was visually doing by looking at the amount of overlap in the Gaussian distributions of the two data sets.

The Student's t-test allows us to do a somewhat similar analysis, but in a more mathematical fashion. Using the Student's t-test we will assign a "percent confidence level" that the two different groups of pennies in fact represent different "data sets" or "populations." To determine with what "percent confidence level" the two sets of data are different, we need to calculate a Student's t-value, which will then be compared with statistical tables. The Student's t-value is calculated according to Eqn. 1A.3

$$t_{calc} = \frac{|\overline{x}_1 - \overline{x}_2|}{s_{pooled}} \sqrt{\frac{n_1 n_2}{n_1 + n_2}} \qquad (1A.3)$$

where

$\overline{x}_1$ = mean value of first set of pennies

$\overline{x}_2$ = mean value of second set of pennies

n_1 = number of pennies in first set, *viz.*, 5

n_2 = number of pennies in second set, *viz.*, 5

s_{pooled} = pooled standard deviation.

The value of s_{pooled} is calculated using the standard deviations of the two sets of pennies, s_1 and s_2, respectively, according to Eqn. 1A.4

$$s_{pooled} = \sqrt{\frac{s_1^2(n_1 - 1) + s_2^2(n_2 - 1)}{n_1 + n_2 - 2}}$$

(1A.4)

Note that t_{calc} contains the same information that we used in our intuitive analysis, how close the two average values were to each other ($\bar{x}_1 - \bar{x}_2$), and the amount of scatter in the data sets (s_{pooled}).

Table 1A.1 contains the theoretical values of Student's t at various confidence levels for your experiment. The table applies only for the number of measurements made here, namely for comparing two sets of values, with each data set containing five data points. If the value of t_{calc} that you calculate based on your experimental data exceeds the value given in Table 1A.1, then there is a difference in the two data sets considered for the level of confidence that you were testing at. If t_{calc} does not exceed the tabulated value in Table 1A.1, then there is not a significant difference in the two data sets. In science we typically use a fairly high level of confidence. Typically, a 95 % confidence level or greater is used. In this experiment there is really only two possible outcomes, either the data sets are significantly different or they are not significantly different. One would have a 50-50 chance of guessing the correct result. We want to be more confident in our answer than this. What does a conclusion at the 95 % confidence level mean? Suppose that we concluded at the 95 % confidence level that the mass of pre-1982 and post-1982 pennies was significantly different. What this would mean is that if we repeated the exact same experiment 100 more times, that 95 times we would get the same result.

Table 1A.1. Student's t for comparing two data sets ($n_1 = n_2 = 5$)

Confidence level %	Student's t
50	0.706
90	1.860
95	2.306
98	2.896
99	3.355
99.5	3.832
99.9	5.041

A more complete listing of Student t-values can be found in standard analytical chemistry textbooks (see for example, *Fundamentals of Analytical Chemistry*, 7th edition, by D. A. Skoog, D. M. West and F. J. Holler, Saunders College Publishing, New York, NY (1996); *Quantitative Chemical Analysis*, 6th edition, by Daniel C. Harris, W. H. Freeman and Company (2003)), standard statistics books (for example, *Introduction to the Theory of Statistics*, 3rd edition, A. M. Mood, F. A. Grayhill and D. C. Boes, McGraw-Hill, Inc., New York, NY (1974); *Methods of Statistical Analysis*, 2nd edition, C. H. Goulden, Wiley, New York, NY (1956)) and in science and engineering handbooks (see for example *Lange's Handbook of Chemistry*, 13th edition, by J. A. Dean, McGraw-Hill, New York, NY (1985); *CRC Standard Mathematical Tables*, 28th edition, W. H. Beyer, CRC Press, Boca Raton, FL (1987)).

NOTE TO INSTRUCTOR – The laboratory manual was written with the intention that students would perform either Experiment 1A or Experiment 1B. The two experiments involve the same statistical treatment, and there is really no need to do both. Experiment 1A requires samples of pre-1982 and post-1982 pennies, which may not be readily available. Experiment 1B compares the density of CocaCola versus Diet Coke.

DATA SHEET – EXPERIMENT 1A

Name: _____

Standard Deviation Determination for 20 post-1982 pennies:

Mass of individual pennies:

_____	_____	_____	_____
_____	_____	_____	_____
_____	_____	_____	_____
_____	_____	_____	_____
_____	_____	_____	_____

Calculated mean of 20 pennies: _____

Calculated standard deviation of 20 pennies: _____

For the pre-1982 pennies:

Penny	Mass (grams)	$x_i - \bar{x}$	$(x_i - \bar{x})^2$
1	_____	_____	_____
2	_____	_____	_____
3	_____	_____	_____
4	_____	_____	_____
5	_____	_____	_____

Calculate the mean ($\Sigma\, x_i\, /\, n$):   _____

Fill in the rest of table above, and calculate ($\Sigma\, (x_i - \bar{x})^2$):   _____

Calculate standard deviation, s: _____

9

For the post-1982 pennies:

Penny	Mass (grams)	$x_i - \bar{x}$	$(x_i - \bar{x})^2$
1	_____	_____	_____
2	_____	_____	_____
3	_____	_____	_____
4	_____	_____	_____
5	_____	_____	_____

Calculate the mean ($\Sigma\, x_i / n$): _____

Fill in the rest of table above, and calculate ($\Sigma\, (x_i - \bar{x})^2$): _____

Calculate standard deviation, s: _____

For the 1943 steel pennies:

Penny	Mass (grams)	$x_i - \bar{x}$	$(x_i - \bar{x})^2$
1	2.6885	_____	_____
2	2.6933	_____	_____
3	2.7036	_____	_____
4	2.6917	_____	_____
5	2.6849	_____	_____

Calculate the mean ($\Sigma\, x_i / n$): _____

Fill in the rest of table above, and calculate ($\Sigma\, (x_i - \bar{x})^2$): _____

Calculate standard deviation, s: _____

Comparison of pre-1982 and post-1982 pennies:

What is the numerical value of s_{pooled}? _____

What is the numerical value of t_{calc}? _____

Statistically, two data sets are "considered different" if they are different at the 95 % confidence level or greater. Does your experimental data fulfill this criteria? In other words, does your value of t_{calc} exceed the value Table 1A.1 (t = 2.306)? _____

Comparison of post-1982 and 1943 steel pennies:

What is the numerical value of s_{pooled}? _____

What is the numerical value of t_{calc}? _____

Statistically, two data sets are "considered different" if they are different at the 95 % confidence level or greater. Does your experimental data fulfill this criteria? In other words, does your value of t_{calc} exceed the value Table 1A.1 (t = 2.306)? _____

Comparison of pre-1982 and 1943 steel pennies:

What is the numerical value of s_{pooled}? _____

What is the numerical value of t_{calc}? _____

Statistically, two data sets are "considered different" if they are different at the 95 % confidence level or greater. Does your experimental data fulfill this criteria? In other words, does your value of t_{calc} exceed the value Table 1A.1 (t = 2.306)? _____

EXPERIMENT 1B: STATISTICAL ANALYSIS OF THE DENSITY OF COCACOLA VERSUS DIET COKE

INTRODUCTION

Aspartame (L-alpha-aspartyl-L-phenylalamine methyl ester; better known as NutraSweet) is a low calorie artificial sweetner used in a wide variety of low- and reduced-calorie foods and beverages. Aspartame is about 160 times sweeter than sucrose in an aqueous solution. In other words, it takes 160 times more sucrose than aspartame to obtain the same level of sweetness. Today's laboratory experiment will examine whether or not the replacement of high fructose corn syrup (used to sweeten CocaCola) with aspartame results in a statistically significant difference in the density of CocaCola of versus Diet Coke. Density is an intensive physical property, and is defined as

$$density = \frac{mass\ of\ object}{volume\ of\ object} \tag{1B.1}$$

as the mass of an object divided by its density.

Scientific experiments are almost always performed in replicated to ensure that the measured values and observations are reproducible. When replicate measurements are peformed one reports not only the individual values, but also the mean value (also referred to as the average value) and standard deviation. The mean value, $\bar{x}$, is obtained by adding all of the individual measurements and then dividing by the total number of measurements.

$$\bar{x} = \Sigma\, x_i\, /\, n \tag{1B.2}$$

where Σ represents the mathematical summation operation, and n is the number of replicate measurements performed. The standard deviation is calculated as

$$s = \sqrt{\frac{\sum(x_i - \bar{x})^2}{n-1}} \tag{1B.3}$$

Standard deviations are generally expressed with a "±" sign; *e.g.* "100.53 ± 0.32". The smaller the standard deviation, the less scattered the numerical values. When the standard deviation is small, the individual values are grouped closer about the mean value.

The calculated standard deviation does have statistical significance in that random experimental errors follow a Gaussian error distribution (*e.g.*, "bell-shaped distribution"). If a given measurement is performed a large number of times, one will obtain a range of numerical values that will be distributed according to a Gaussian error distribution, provided that one does not make systematic errors of measurement. For random errors, small errors are much more probable than large errors, and positive deviations are just as likely as negative ones. In the laboratory, we rarely repeat the same experiment enough times to truly obtain a Gaussian distribution. We can take our smaller set of measured values; however, calculate the mean and standard deviation for our smaller data set, and using statistics project what the Gaussian distribution would look like, had we performed the much larger number of experimental measurements. The standard deviation is used to describe the variation in a finite data set, whereas the variance (denoted as σ), or population standard deviation as it is sometimes called, is used in an infinite population. The variance is calculated in a similar fashion as the standard deviation, except that the denominator is n, rather than n-1 (see Equation 1B.3).

The population standard deviation measures the width of the Gaussian distribution curve. The larger the value of σ, the broader is the curve. For any Gaussian distribution, 68.3 % of the area under the curve falls in the range of $\bar{x}$ ± 1σ. That is, more than two-thirds of the measurements are expected to lie within one standard deviation of the mean. Also, 95.5 % of the area under the Gaussian curve lies within $\bar{x}$ ± 2σ, and 99.7 % of the area lies within $\bar{x}$ ± 3σ. (See Figure 1B.1) Only a very, very small percentage of the experimental measurements would be expected to be more than 3 standard deviations from the mean as the result of random errors. One can use this rationalization in reverse. Suppose that one did measure an experimental value that was more than three standard deviations larger than the calculated average value for five replicate experimental determinations. The experiment had been performed five times, and one value was significantly different from the other values. What would you think or conclude based on the preceding discussion? That the difference between the one apparent "outlier" and the other four values resulted from solely random errors, or that one unknowingly made a mistake in that particular experimental measurement?

In Experiment 1B we will determine the average density and standard deviation of a sample of CocaCola and Diet Coke, and using statistics determine whether or not there is a significant difference in the density of the two beverage formulations. As noted earlier, Diet Coke contains aspartame, which is 160 times sweeter than sucrose in aqueous solution. An indication that there is less sweetener in Diet Coke than in

CocaCola is the observation that aspartame is listed as the third most abundant ingredient in Diet Coke (after water and caramel), whereas high fructose corn syrup is listed as the second largest ingredient in CocaCola.

GAUSSIAN DISTRIBUTION

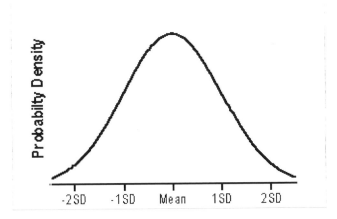

Figure 1B.1. Cumulative probability distribution of a Gaussian bell-shaped distribution. The standard probability distribution is scaled so that the area under the curve equals 1.0.

DENSITY DETERMINATIONS

In order to same time you will be working groups of four. One group of two students will determine the density of CocaCola, while the other group measures the density of Diet Coke. After all of the densities are recorded, the two groups will exchange data so that each student has the entire set of experimental data recorded on his/her Laboratory Data Sheet. **Each student must turn in a completed Data Sheet to the Teaching Assistant to grade**.

Pour about 125 ml of the soft drink that you will be studying into a clean, dry 250-ml beaker. [**Note to Instructor** – At least 24 hours before the experiment is to be performed, samples of CocaCola and Diet Coke are emptied into separate large Erlenmeyer flasks each equipped with a magnetic stir bar and watch glass. The solutions are stirred for one hour to remove the dissolved carbon dioxide. Approximately 1 Liter of each soda is required for each 24-student lab section. CocaCola cannot be prepared more than two days in advance of the experiment or bacteria will start to grow in the degassed high fructose corn syrup solution (sugar) solution.] Next, you will need to weigh an empty 125-ml Erlenmeyer flask on an electronic top-loading balance. Record the mass of the empty beaker to three decimal places on your Laboratory Data Sheet for Experiment 1B. Be sure to record the mass on the appropriate line. Very carefully add by pipette 25 ml of CocaCola (or Diet Coke) to the flask, and reweigh. Record the mass of the empty flask + CocaCola (or Diet Coke) after first 25-ml aliquot on the Data Sheet. The mass of the Coke (or Diet Coke) that was added to the flask is now calculated by difference. The density of the CocaCola is calculated by substituting the mass of the sample and volume of the sample into Eqn. 1B.1. **DO NOT EMPTY THE FLASK AS THE FINAL WEIGHT WILL SERVE AS THE INITIAL WEIGHT FOR THE NEXT 25-ml ALIQUOT ADDITION.** Pipet a second 25-ml portion of CocaCola (or Diet Coke) into the flask, and reweigh. Record the new weight on the Data Sheet in the space provided for "mass of flask + CocaCola (Diet Coke) after second 25-ml aliquot." The mass of the second 25-ml of CocaCola (Diet Coke) is calculated by subtracting the numerical value recorded on line number 2 (or line number 15 in the case of Diet Coke) from the value on line number 5 (or line number 18 in the case of Diet Coke.) Repeat the measurements for the third and fourth 25-ml aliquot of CocaCola (or Diet Coke).

After all of the densities have been measured exchange data with your other two laboratory partners so that everyone has the entire set of experimental data for both CocaCola and Diet Coke recorded on his/her Laboratory Data Sheet. It is now time to calculate the average density and standard deviation of each soft drink. Most scientific hand-held calculators have a built-in statistic package that will perform this calculation. If you know how to use this function on your calculator, fine. Go ahead and calculate both numerical values and record the mean and standard deviation on the Data Sheet in the indicated space.

The Pasco system has the capability of performing statistical calculations. To calculate and average and standard deviation using the built-in software, double on Data Studio. You have now opened the system. **Double click on the activity titled Enter Data.** What should now appear on the screen is a table on the left-hand side, and a graph on the right-hand side. Enter the values to be averaged in the x-column of the table. Don't worry about what is happening with the y-column. Once all of the values are entered, **click on the Σ key on the Table toolbar.** The mean should now be displayed at the bottom of the column. If you want to calculate the standard deviation, go back up to the top of the table, and **click on the ▼ arrow by the Σ key,** scroll down to standard deviation, and hit enter. The standard deviation is now at the bottom the page, along with the average. When you get ready to enter the next set of data, you can either enter the next set right over the previous data set (be sure to delete out any extra data points), or you can clear the numerical values in the table by **mouse clicking on New Data** on the top tool bar. [Alternatively, you can bring down an empty data table by **mouse clicking on the Experiment key** on the top toolbar, and then **scrolling down to New Empty Data Table**. The latter method will probably take a few minutes longer. Since you have to type all of the numbers anyway, it is probably easier to type over existing numbers, than bringing down new tables and then typing the new numbers into an empty table.) Now perform the calculations long-hand for each soft drink by summing up the individual $(x_i - \bar{x})^2$ as directed on the Data Sheet. The averages and standard deviations that you calculated long-hand should agree with the values calculated with the Pasco software. If not, check over the calculations until you find the error.

Microsoft Excel Instructions for Average and Standard Deviation Calculations: Some of the calculations during the semester will have to be performed outside of the laboratory time. This will be true of the average and standard deviation calculations based on the experimental data from the entire class. You can perform the calculation using Microsoft Excel spreadsheet. The general computer access laboratories across campus should have Microsoft Excel as part of the standard Microsoft Windows bundle of software. The Excel spreadsheet will appear as a grid of cells, which are identified by both a column alphabetical letter and row numerical value. In cells A5 through A8 enter the density of the four 25-ml aliquot samples of Coke. In cell A3 type: **=AVERAGE(A5:A8)** . Hit **ENTER** -By typing this command you are programming the computer to calculate the average value of the numbers that are entered in cells A5 through A8. The calculated mean (average value) should now appear in cell A3. To calculate the standard deviation of the numerical values stored in cells A5 through A8, go to cell B3 and type: **=STDEV(A5:A24).** Hit **ENTER** - You have just instructed the computer to calculate the standard deviation of the values that are entered in cells A5 through A8. The standard deviation should now appear in cell B3. Microsoft Excel has several built-in statistical features that we may need to access during the semester to complete parts of the laboratory experiments.

COMPARISON OF DIFFERENT DATA SETS USING STUDENT'S t-TEST

Next we need to analyze the experimental results using statistics to determine how significant the difference is between the density of CocaCola versus the density of Diet Coke. Some idea of whether or not there is a significant difference between any two data sets can be gained simply by looking at the data sets being compared. Suppose that the two data sets had exactly the same average value to the third decimal place. What would your intuition tell you? That the two data sets were the same, at least as far as the mass is concerned. It is very unlikely though, that the two data sets will have exactly the same average value. Now what type of information would you want to have? You would probably want to know not only the average mass of each data set, but also how much variation was there in the individual masses for each data set. In other words, you would want to know the mean and standard deviation of both data sets being compared. As was stated earlier, we could use the calculated mean and standard deviation for each finite data set to construct a Gaussian distribution for the density of CocaCola and for the density of Diet Coke using statistical methods. If I were asked to compare the density of CocaCola versus the density of Diet Coke, I could look at the amount of overlap between the two respective Gaussian distributions. For example, if the two Gaussian distributions were super-imposable on one another, then I would likely state that the two soft drinks are definitely identical, at least as far their density is concerned. If there was a lot of overlap in the two Gaussian distributions, I would likely state that I believe that the two soft drinks are identical. A little bit less overlap, that the two beverage densities may be identical. Even less overlap, that the two densities are likely different. And finally to the point where there is no overlap between the two data sets, where I would likely conclude that the two data sets are indeed different. You will note that I was hedging my statement depending on how confident I was in the conclusion that I was making based on the analyses that I was visually doing by looking at the amount of overlap in the Gaussian distributions of the two data sets.

The Student's t-test allows us to do a somewhat similar analysis, but in a more mathematical fashion. Using the Student's t-test we will assign a "percent confidence level" that the two soft drinks in fact represent different "data sets" or "populations." To determine with what "percent confidence level" the two sets of data are different, we need to calculate a Student's t-value, which will then be compared with statistical tables. The Student's t-value is calculated according to Eqn. 1B.4

$$ t_{calc} = \frac{\left| \bar{x}_1 - \bar{x}_2 \right|}{s_{pooled}} \sqrt{\frac{n_1 n_2}{n_1 + n_2}} \tag{1B.4} $$

where

$\bar{x}_1$ = mean value of CocaCola density

$\bar{x}_2$ = mean value of Diet Coke density

n_1 = number of times the density of CocaCola was measured, *viz.*, 4

n_2 = number of times the density of Diet Coke was measured, *viz.*, 4

s_{pooled} = pooled standard deviation.

The value of s_{pooled} is calculated using the standard deviations of the two sets of densities, s_1 and s_2, respectively, according to Eqn. 1B.5

$$s_{pooled} = \sqrt{\frac{s_1^2(n_1-1) + s_2^2(n_2-1)}{n_1 + n_2 - 2}}$$ (1B.5)

Note that t_{calc} contains the same information that we used in our intuitive analysis, how close the two average values were to each other ($\bar{x_1} - \bar{x_2}$), and the amount of scatter in the data sets (s_{pooled}).

Table 1B.1 contains the theoretical values of Student's t at various confidence levels for your experiment. The table applies only for the number of measurements made here, namely for comparing two sets of values, with each data set containing four data points. If the value of t_{calc} that you calculate based on your experimental data exceeds the value given in Table 1B.1, then there is a difference in the two data sets considered for the level of confidence that you were testing at. If t_{calc} does not exceed the tabulated value in Table 1B.1, then there is not a significant difference in the two data sets. In science we typically use a fairly high level of confidence. Typically, a 95 % confidence level or greater is used. In this experiment there is really only two possible outcomes, either the data sets are significantly different or they are not significantly different. One would have a 50-50 chance of guessing the correct result. We want to be more confident in our answer than this. What does a conclusion at the 95 % confidence level mean? Suppose that we concluded at the 95 % confidence level that the density of Coke and Diet Coke was significantly different. What this would mean is that if we repeated the exact same experiment 100 more times, that 95 times we would get the same result.

Table 1B.1. Student's t for comparing two data sets ($n_1 = n_2 = 4$)

Confidence level %	Student's t
50	0.718
90	1.943
95	2.447
98	3.143
99	3.707
99.5	4.317
99.9	5.959

A more complete listing of Student t-values can be found in standard analytical chemistry textbooks (see for example, *Fundamentals of Analytical Chemistry*, 7th edition, by D. A. Skoog, D. M. West and F. J. Holler, Saunders College Publishing, New York, NY (1996); *Quantitative Chemical Analysis*, 6th edition, by Daniel C. Harris, W. H. Freeman and Company (2003)), standard statistics books (for example, *Introduction to the Theory of Statistics*, 3rd edition, A. M. Mood, F. A. Grayhill and D. C. Boes, McGraw-Hill, Inc., New York, NY (1974); *Methods of Statistical Analysis*, 2nd edition, C. H. Goulden, Wiley, New York, NY (1956)) and in science and engineering handbooks (see for example *Lange's Handbook of Chemistry*, 13th edition, by J. A. Dean, McGraw-Hill, New York, NY (1985); *CRC Standard Mathematical Tables*, 28th edition, W. H. Beyer, CRC Press, Boca Raton, FL (1987)).

NOTE TO INSTRUCTOR – The laboratory manual was written with the intention that students would perform either Experiment 1A or Experiment 1B. The two experiments involve the same statistical treatment, and there is really no need to do both. Experiment 1A requires samples of pre-1982 and post-1982 pennies, which may not be readily available. Experiment 1B compares the density of CocaCola versus Diet Coke.

DATA SHEET – EXPERIMENT 1B

| Coca Cola | vs | Diet Coke |

Name: _____

Density Measurements on CocaCola:

1. Mass of empty Erlenmeyer flask, g: _____
2. Mass of flask + CocaCola after first 25-ml aliquot, g: _____
3. Mass of first 25-ml aliquot of CocaCola, g: _____
4. Density of first 25-ml aliquot of CocaCola, g/ml: _____
5. Mass of flask + CocaCola after second 25-ml aliquot, g: _____
6. Mass of second 25-ml aliquot of CocaCola, g: _____
7. Density of second 25-ml aliquot of CocaCola, g/ml: _____
8. Mass of flask + CocaCola after third 25-ml aliquot, g: _____
9. Mass of third 25-ml aliquot of CocaCola, g: _____
10. Density of third 25-ml aliquot of CocaCola, g/ml: _____
11. Mass of flask + CocaCola after fourth 25-ml aliquot, g: _____
12. Mass of fourth 25-ml aliquot of CocaCola, g: _____
13. Density of fourth 25-ml aliquot of CocaCola, g/ml: _____

Density Measurements on Diet Coke:

14. Mass of empty Erlenmeyer flask, g: _____
15. Mass of flask + Diet Coke after first 25-ml aliquot, g: _____
16. Mass of first 25-ml aliquot of Diet Coke, g: _____
17. Density of first 25-ml aliquot of Diet Coke, g/ml: _____
18. Mass of flask + Diet Coke after second 25-ml aliquot, g: _____
19. Mass of second 25-ml aliquot of Diet Coke, g: _____
20. Density of second 25-ml aliquot of Diet Coke, g/ml: _____
21. Mass of flask + Diet Coke after third 25-ml aliquot, g: _____
22. Mass of third 25-ml aliquot of Diet Coke, g: _____
23. Density of third 25-ml aliquot of Diet Coke, g/ml: _____
24. Mass of flask + Diet Coke after fourth 25-ml aliquot, g: _____
25. Mass of fourth 25-ml aliquot of Diet Coke, g: _____
26. Density of fourth 25-ml aliquot of Diet Coke, g/ml: _____

For Density of CocaCola:

Aliquot	Density (g/ml)	$x_i - \bar{x}$	$(x_i - \bar{x})^2$
1	_____	_____	_____
2	_____	_____	_____
3	_____	_____	_____
4	_____	_____	_____

Calculate the mean ($\Sigma\, x_i\, /\, n$): _____

Fill in the rest of table above, and calculate ($\Sigma\, (x_i - \bar{x})^2$): _____

Calculate standard deviation, s: _____

For Density of Diet Coke:

Aliquot	Density (g/ml)	$x_i - \bar{x}$	$(x_i - \bar{x})^2$
1	_____	_____	_____
2	_____	_____	_____
3	_____	_____	_____
4	_____	_____	_____

Calculate the mean ($\Sigma\, x_i\, /\, n$): _____

Fill in the rest of table above, and calculate ($\Sigma\, (x_i - \bar{x})^2$): _____

Calculate standard deviation, s: _____

Comparison of Density of CocaCola Versus Diet Coke:

What is the numerical value of s_{pooled}? _____

What is the numerical value of t_{calc}? _____

Statistically, two data sets are "considered different" if they are different at the 95 % confidence level or greater. Does your experimental data fulfill this criteria? In other words, does your value of t_{calc} exceed the value Table 1B.1 (t = 2.447)? _____

EXPERIMENT 2: PHYSICAL AND CHEMICAL PROPERTIES LIQUID COMPOUNDS

INTRODUCTION

Matter is defined as anything that has mass and occupies space. Everything that you see, and a large number of items that you cannot see, is matter. Four fundamental ways in which matter can be classified and described are according to its physical state (such as solid, liquid, gas or liquid crystal), according to its composition (as an element, compound, or mixture), according to its physical properties (density, melting point temperature, boiling point temperature, viscosity, color, odor, *etc.*) or according to its chemical properties (flammability, reactivity with sodium metal, *etc.*). Physical properties can be measured without changing the identity and composition of the substance. For example, in determining the melting point temperature of water, one measures the temperature of the solid-to-liquid transition. Ice is $H_2O_{(solid)}$, and during the measurement of the melting point, the molecule is the same, namely H_2O. No chemical bonds are broken or formed during the measurement. Flammability, on the other hand, is a chemical property. In determining whether or not a compound is flammable it may be converted to an oxide, such as in the combustion of butane gas

$$C_4H_{10(gas)} + 6.5\ O_{2(gas)}\ \text{-----}>\ 4\ CO_{2(gas)} + 5\ H_2O_{(gas)}$$

or oxidation of metallic iron

$$2\ Fe_{(solid)} + 1.5\ O_{2(gas)}\ \text{-----}>\ Fe_2O_{3(solid)}$$

to form ferric oxide (rust).

In today's laboratory experiment you will examine a variety of physical and chemical properties of two liquids. By comparing your observations to a list of properties of "known" substances, you should be able to identify both substances. Some of the properties that will be studied are "qualitative" in nature, whereas other properties are more quantitative. Qualitative observations are often more subjective than are quantitative measurements that have a numerical value associated with the property. Quantitative measurements are more informative than qualitative observations. For example, water and chloroform are both liquids at room temperature. Observing that a substance is liquid at room temperature would not enable one to distinguish between water and chloroform. ("Liquid" and "solid" at room temperature is essentially a qualitative observation concerning the melting/freezing point temperature.) If one were to have a more quantitative measurement for the solid-to-liquid phase transition, then one could distinguish between water and chloroform. The melting/freezing point temperature of water (T = 0 °C) and chloroform (T = - 63 °C) are sufficiently different that one should be able to tell which chemical is which.

Solubility is another one of the physical properties that can be described in both a qualitative and quantitative manner, particularly in the case of liquid substances. Pairs of liquids such as ethyl alcohol and water that mix in all proportions are completely miscible, whereas liquids that do not dissolve appreciably in one another are said to be immiscible with each other. Strictly speaking, no two liquids are completely immiscible as all substances are at least partially soluble in all other substances. The solubility may be extremely small; however, so small that one cannot visually tell that any of the first liquid dissolved in the second liquid, and vice versa. On a more quantitative basis, the solubility would be given in terms of the actual mass (or volume) of solute that dissolved in a known mass (or volume) of solvent. Quantitative measurements are more informative than qualitative observations. For today's laboratory experiment the qualitative observations of "miscible" and "immiscible" will allow you narrow down the list of possibilities.

Words like "oily", "viscous", "syrupy" and "mobile" are qualitative observations concerning the substances viscosity. Viscosity is defined as the resistance of a liquid to flow, and as you will later discover in the accompanying general chemistry lecture course, viscosity is related to the intermolecular forces between neighboring molecules. The observation can be made a bit more quantitative by measuring the amount of time required for the liquid substance to flow a fixed distance, or by measuring the amount of time required for a fixed volume of liquid to drain from lets say a 1 ml graduated pipette. [Note to Instructors – Several published papers describe methods for making an inexpensive viscometer suitable for student use from common, readily available glassware. For example, see L. G. Draignault, D. C. Jackman and D. P. Rillema, *J. Chem. Educ.*, **67**, 81-82 (1990); T. J. McCullough, *J. Chem. Educ.*, **61**, 68-69 (1984); and J. R. Khurma, *J. Chem. Educ.*, **68**, 63 (1991).]

Properties that are obtained through human sensory detection, by their very nature, do have a certain level of ambiguity. The way that one person registers the "color" or "smell" of a particular substance may be completely different than how someone else's senses register these two physical properties. You know that some individuals may have the visual impairment of colorblindness. Even if a person is not diagnosed as being colorblind, he/she may still have difficulty distinguishing between red and green. This is true for the other colors as well. Everyone draws upon his or her past experiences. One individual may think that a piece of paper is "golden rod" in color, while another person, looking at the exact same piece of paper, may write down that the paper is "yellowish-orange" (perhaps orangish-yellow) in color. A third person reading what the other two individuals wrote down, "golden rod" versus "yellowish-orange", would not necessarily know, unless otherwise told, that both individuals had looked at exactly the same piece of paper. We all draw upon our past experiences in interpreting our sensory responses. One additional note – colors can vary if even a slight amount of impurity is present. A pure sample might be clear, but appear yellow when impurities are present. The take-home message is that while color and smell are useful physical properties to know, compound identification must not be based solely on color and smell.

EXPERIMENTAL PHYSICAL PROPERTY SENSORY AND APPEARANCE OBSERVATIONS

Before you make any of the quantitative experimental measurements, try to narrow down the list of possibilities through some of the more qualitative observations. Obtain about 20 ml of each of the unknowns A, B and C in different 100 ml beakers. This should be enough of each unknown for doing of all today's laboratory measurements. Describe the color and fluidity of each unknown on the laboratory Data Sheet. Now raise the beaker containing unknown A level with, and about three or four inches away from your nose. Wave your hand over the top of the beaker to waft the order toward you. Describe any odor. Repeat the process for unknowns B and C.

The solubility is to be studied next. Transfer by pipet about 2 ml of unknown A into two separate test tubes. Each test tube should contain 2 ml of unknown A. To the first test tube, carefully add about 2 ml of deionized water. Do the liquids mix or form two layers. If they mix, then record "miscible" on the laboratory Data Sheet, otherwise, record "immiscible". To the second test tube, add 2 ml of cyclohexane. Record the solubility of unknown A and either "miscible" or "immiscible" on the Data Sheet. Repeat the solubility observations for unknowns B and C. [Note – A third solvent that could be added to the miscibility determinations is methanol. Like water, methanol is a fairly polar molecule. All of the compounds listed in Table 2.1 would be miscible with methanol, except for hexane, isooctane and hexadecane.]

As the final qualitative physical property observation, you will observe whether of not the substances freezes in an ice-water bath, and whether or not the substance freezes in an ethanol-dry ice bath (or acetone-dry ice bath). A more quantitative study of the liquid-to-solid phase transition will be made in a later experiment when you determine the freezing point temperatures of various "pure" substances and liquid mixtures using a cooling curve method. For now, you will simply observe whether the substance is a solid or liquid at the two bath temperatures: 0 °C for the ice-water bath; –78 °C for dry ice; and about -63 °C for an acetone-dry bath and –71 °C for ethanol-dry ice bath. In performing the freezing point observation be aware that as a liquid is cooled below its freezing point, it may supercool, that is, temporarily remain a liquid even though it is below the freezing point. A liquid generally though does not supercool more than 10 degrees below its freezing point temperature. A substance like carbon tetrachloride (T_{mp} = -23 °C) should readily freeze at dry ice temperatures; however, a substance such as chloroform (T_{mp} = -63 °C) may supercool.

CAUTION – DRY ICE CAN CAUSE SEVERE FROST BITE AND PERMANENT PHYSICAL INJURY – UNDER NO CIRCUMSTANCES SHOULD YOU POKE YOUR FINGERS INTO THE DRY ICE BATH. ANY PHYSICAL CONTACT BETWEEN YOURSELF AND THE DRY ICE MUST BE KEPT TO A MINIMUM – ONLY A SECOND OR TWO. Cloth gloves are excellent protection for handling dry ice, or alternatively, paper towels can be used to wrap around pieces of dry ice when handling and when inserting pieces of dry ice into the alcohol bath.

Also, as you will discover in Experiment 8, impurities in a chemical substance causes the freezing point temperature to be lowered. Place a few drops of each unknown in a separate test tube, and record your observations regarding whether or not the substance solidifies at the ice-water bath temperature and at the dry ice bath temperature. It should take only a few minutes of for the water bath, and only a few seconds for the dry ice bath for the substance to solidify (or start to solidify). For each unknown, you should be able now to eliminate several of the compounds listed in Table 1.

EXPERIMENTAL PHYSICAL PROPERTY MEASUREMENTS

Density will be the first of two physical properties that will be quantitatively measured. Place a dry test tube inside a clean, dry 150-ml beaker and weigh the beaker + test tube on a top loading electronic balance. Record the mass on the laboratory Data Sheet in the section titled "Physical Property Quantitative Measurements". Next pipet an exact volume of unknown A (about 2 ml) into the test tube, and record the volume on the Data Sheet. For pipets, graduated cylinders and burets, the liquid level is read from the bottom of the U-shaped meniscus (See Figure 2.1). Reweigh the beaker + test tube + unknown in the test tube. You will need to be prompt in weighing once the liquid is added. Volatile liquids like acetone, chloroform and dichloromethane do evaporate fairly fast, and if one is not careful, the recorded mass may be erroneously low due to evaporation. Record the mass of the beaker + test tube + unknown on your Data Sheet. The mass of the liquid that was pipetted is calculated by difference. Calculate the density of unknown A by substituting the measured mass (in grams) and volume (in ml) into Eqn. 2.1 below:

density = mass of substance/volume of substances (2.1)

Repeat the procedure for unknowns B and C.

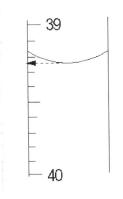

Figure 2.1. Determination of volumetric reading for a pipet, buret or graduated-cylinder. For a clear liquid, the volumetric reading is made from the bottom of the U-shaped meniscus. One should always try to "guesstimate" the reading to one additional decimal. In the example to the left, the volumetric glassware is marked off to a tenth of a milliliter. The recorded volume should then be to the hundredths of a milliliter. One should be able to guesstimate whether the meniscus is halfway between two lines, a quarter of the way between two lines, etc.

From an experimental standpoint, the volume measurement will likely have a larger experimental uncertainty than the mass. The mass of 2 ml of the unknown should be somewhere between 1.5 grams and 3.0 grams, depending on what unknown you have. The manufacturer's specification for the electronic balances is ±2 in the last decimal place, which for a balance capable of measuring to the milligram level (third decimal place) corresponds to ±0.002 grams. A quick, "worse case" estimate of the uncertainty in the mass could be obtained by assuming that the mass of the empty beaker was "truly" ±0.002 grams larger than what was displayed on the balance and that the mass of the beaker + liquid was "truly" ±0.002 smaller than what the balance read. The "worst case" estimate for the uncertainty in the mass of the liquid would be ±0.004 grams. The most probable error, based on a statistical technique called "error propagation" is slightly smaller, e.g., ±0.003. In either case, the error is small compared to the 1.5 to 3.0 gram mass itself. The uncertainty in the pipet volume is a little bit harder to calculate because there are so many different types of pipets that one could use to transfer 2 ml of unknown, and each type of pipet has a different tolerance. For a Class A 2-ml pipet (if the pipet is Class A you will see Class A and the tolerance written on the stem of the pipet), the tolerance is ±0.006 ml. You will notice that the uncertainty/error in volume is at double what the uncertainty was in the case the mass of the liquid. A Class B 2-ml pipet would have a tolerance of ±0.012 ml. The uncertainty/error for a graduated pipet is even larger, and is in the range of ±0.02 to ±0.05 ml, depending on how the pipet markings are graduated. The manufacturer's specifications assume that one is using the pipet correctly, and lining the bottom of the U-shaped meniscus up with the line. An error propagation calculation of the most "probable error" in the calculated density requires knowledge of calculus, and is beyond the scope of the introductory general chemistry laboratory course; however, for beginning students a error of ±0.01 to ±0.02

grams/cm^3 would not be all that bad, remembering that for many students this will be the first time that you have used this particular type of chemical glassware.

The second of two physical properties that will be quantitatively measured is the boiling point temperature, T_{bp}. Assemble a heating system by attaching an O-ring clamp onto a ring stand. Leave sufficient room below the O-ring clamp for placing a Bunsen burner. Place a wire gauze on the O-ring clamp. Half-fill a 250-ml beaker with water, add a couple of boiling chips (Tamer Tabs, Boileasers, *etc.*), and place the beaker on the wire gauze. The boiling will prevent large bubble formation and "bumping" of the liquid, which can happen if there is no place for small bubbles to from around.

Pipet about 2 ml of unknown A into a test tube and clamp the test tube inside the beaker so that the liquid level of the unknown in the test tube is below the water level in the 250-ml beaker. Add a boiling chip or splint to the unknown liquid inside the test tube. The boiling chip or splint will ensure the formation of small bubbles. Connect the rubber tubing of the Bunsen burner to the natural gas outlet in along the laboratory bench. Close the air holes at the base of the Bunsen burner. Strike a match and open the gas tap slightly. Hold the lit match over the top of the Bunsen burner to ignite the flame (be sure to keep your hand off to the side so as not to get burned). Adjust the air holes until the flame is pale blue.

Begin heating the water from below. Watch the unknown liquid for signs of boiling. When the liquid begins to boil, measure the temperature by placing a thermometer into the unknown liquid. Be sure that the liquid is boiling. The temperature of a slowly boiling liquid (called simmering" and a vigorously boiling liquid are the same. Do not mistake the effervescing of dissolved gases as "simmering". Vigorous boiling will let you know for sure. Once the temperature of the boiling liquid is measured, record the value of T_{bp} on the laboratory Data Sheet. The maximum temperature that can be measured with the technique is 100 °C. If the unknown does not boil then its boiling point temperature must be above 100 °C. On the Data Sheet simply record on the line for the boiling point temperature "above 100 °C'. [Note – you will learn later in the chemistry lecture course that a substance's boiling point temperature depends on the atmospheric pressure. The boiling temperature temperatures, T_{bp}, that are listed in Table 2.1 are for a pressure of 1 atmosphere. At our elevation, the observed values of T_{bp} will be one or degrees less than the literature values. Repeat the procedure for unknowns B and C.

Once all of the physical properties have been made, attempt to identify all three unknowns. Do not spend too much time on the identifications as there are two chemical properties that need to be determined. If you do not get the unknowns quickly identified, come back to this part of the laboratory exercise.

CHEMICAL PROPERTY OBSERVATIONS

The first of two chemical properties that will be determined is whether or not the unknown substances are flammable. Pipet 1 ml of unknown A into an evaporating dish. Light a splint away from the dish. Carefully attempt to ignite the unknown in the evaporating dish. Record you observation on the laboratory Data Sheet in the section labeled "Chemical Property Observations". Repeat the procedure for unknowns B and C.

The last chemical property to be determined is whether or not the unknown substances react with sodium metal. To perform this part of the experiment, put about one inch depth of each unknown A, B and C into separate 150-ml beakers. Take these beakers to the fume hood, and obtain a piece of sodium metal from the Teaching Assistant. Using the tweezers that have been provided, take the piece of sodium and lay it on a paper towel. Dry the sodium and quickly add it to the sample of unknown A. Record your observations on the Data Sheet. Does the sodium metal react with the unknown? Repeat the procedure for unknowns B and C, and record your observations. Return any unused portion of sodium to the Teaching Assistant.

CAUTION – SODIUM METAL REACTS VIOLENTLY WITH WATER – DO NOT TOUCH THE METAL WITH YOUR HANDS OR INGEST!

If you were not able to identify any of your three unknowns, try combining the chemical property observations with the physical property measurements.

Table 2.1. Physical Properties of Select Liquid Substances

Substance	Appearance & Odor	T_{mp} (°C)	T_{bp} (°C)	Density[a] (grams/ml)	Solubility Water	Solubility Cyclohexane
acetophenone	clear, mobile fruity odor, slightly sweet	19	202	1.030	No	Yes
acetone	clear, mobile solvent odor	-94	56	0.791	Yes	Yes
benzaldehyde	clear, oil almond odor	-26	179	1.045	No	Yes
carbon tetrachloride	clear, heavy solvent odor	-23	77	1.594	No	Yes
chloroform	clear, heavy sweet odor	-63	62	1.492	No	Yes
cyclohexane	clear, mobile alkane odor	6	81	0.779	No	Yes
dichloromethane	clear, mobile sweet odor	-95	40	1.325	No	Yes
ethyl acetate	clear, mobile solvent odor	-84	77	0.902	No	Yes
ethyl alcohol	clear, mobile spirit odor	-114	78	0.789	Yes	Yes
ethylene glycol	clear, syrupy no odor	-16	198	1.113	Yes	No

glycerol	clear, syrupy no odor	20	290	1.250	Yes	No
hexadecane	clear, fairly viscous barely detectable odor	18	287	0.773	No	Yes
hexane	clear, mobile alkane odor	-95	69	0.659	No	Yes
isooctane	clear, mobile alkane odor	-107	99	0.692	No	Yes
methyl benzoate	clear, oily pleasant odor	-12	200	1.088	No	Yes
nitrobenzene	off-white color almond odor	6	211	1.196	No	Yes
toluene	clear, refractive benzene-solvent like odor	-95	111	0.865	No	Yes
p-xylene	clear, mobile benzene-like odor	13	138	0.866	No	Yes
water	clear, no odor	0	100	1.000	Yes	No

[a] Densities pertain to 25 °C.

Table 2.2. Chemical Properties of Selected Liquid Substances

Substance	Flammable	Reacts with Sodium
acetophenone	Yes	No
acetone	Yes	No
benzaldehyde	Yes	No
carbon tetrachloride	No	No
chloroform	No	No
cyclohexane	Yes	No
dichloromethane	Yes	No
ethyl acetate	Yes	No
ethyl alcohol	Yes	Slow
ethylene glycol	Yes	Slow
glycerol	Yes	Slow
hexadecane	Yes	No
hexane	Yes	No
isooctane	Yes	No
methyl benzoate	Yes	No
nitrobenzene	Yes	No
toluene	Yes	No
p-xylene	Yes	No
water	No	Yes

DATA SHEET – EXPERIMENT 2

B
A

Immiscible liquids

Name: _____

Physical Property Sensory and Appearance Observations

	Unknown A	Unknown B	Unknown C
1. Appearance (color, fluidity, *etc.*):	_____	_____	_____
2. Odor:	_____	_____	_____
3. Solubility in water, yes or no:	_____	_____	_____
4. Solubility in cyclohexane, yes or no:	_____	_____	_____
5. Freezes in ice-water bath, yes or no:	_____	_____	_____
6. Freezes in dry-ice bath, yes or no:	_____	_____	_____

Physical Property Quantitative Measurements

7. Mass of beaker + test tube, g:	_____	_____	_____
8. Mass of beaker + test tube + liquid unknown, g:	_____	_____	_____
9. Mass of liquid unknown, g:	_____	_____	_____
10. Volume of liquid unknown, ml:	_____	_____	_____
11. Density of unknown, grams/ml:	_____	_____	_____
12. Boiling point temperature, °C:	_____	_____	_____
13. Identity of unknown based on physical properties:	_____	_____	_____

Chemical Property Measurements

14. Flammable, yes or no:	_____	_____	_____
15. Reaction with sodium, yes or no:	_____	_____	_____
16. Unknown based on physical & chemical properties:	_____	_____	_____

35

EXPERIMENT 3A: DETERMINATION OF SIMPLE EMPIRICAL FORMULA AND WATERS OF HYDRATION

INTRODUCTION

An important facet of chemistry is the preparation of new chemical compounds, substances that on a molecular (or nanoscale) involve a new, unique combination and positional arrangement of atoms. New compounds may have physical and chemical properties similar to those of already existing compounds, or their properties may be quite different. All compounds contain at least two different elements, and most contain more than two elements. In a molecular compound, atoms of two or more elements are combined into independent units molecules. A given compound always has the same relative number of and kind of atoms.

Chemists use chemical formulas to denote the number and kinds of atoms that have been combined to form one molecule of the specific compound being described. The simple empirical formula, which will be determined in today's experiment, gives only the relative number of atoms of each element in the molecule. The molecular formula is more informative, in that this formula provides the actual numbers and types of atoms in the molecule. There are several compounds that have a simple empirical formula of CH. Acetylene (C_2H_2), benzene (C_6H_6) and cyclobutadiene (C_4H_4) are three such molecules. One would need the structural formula (also called molecular structure) in order to know how the atoms are attached in the molecule.

The simple empirical formula of a compound can be deduced by measuring the elemental mass percent composition. Acetylene, cyclobutadiene and benzene contain 92.26% C and 7.74% H by mass. Every compound that has an empirical formula of CH has exactly the same mass percentage of carbon and hydrogen. In today's experiment the empirical formula of a compound formed by the combustion of magnesium metal in air

$$x\ Mg_{(solid)} + (y/2)\ O_{2(gas)} \longrightarrow Mg_xO_{y(solid)}$$

will be determined. The combustion will be carefully performed in order to insure that the chemical reaction goes to completion, and that no product is lost. If both conditions are met, then the conservation of mass requires that all of the magnesium that was initially present must end up as Mg_xO_y. The mass oxygen in the product can be calculated from:

mass of oxygen in product = mass of product – mass of Mg used   (3A.1)

The mass percentage of oxygen and magnesium in Mg_xO_y are:

mass % oxygen = 100 × (mass of oxygen in product/mass of product)   (3A.2) mass % magnesium = 100 × (mass of magnesium/mass of product)

(3A.3)

given by Eqns. 3A.2 and 3A.3, respectively. The subscripts "x" and "y" in the chemical formula denote the relative number of atoms of Mg and O in the simple empirical formula, and not the mass percentages that were just calculated.

The number of atoms is proportional to the number of moles, with the proportionality constant being Avogadro's number. It is more convenient to work in mole numbers rather than atom numbers. The relative mole numbers of Mg and O are computed by

moles of Mg = mass % of Mg/molar mass of Mg (3A.4)

moles of O = mass % of O/molar mass of O (3A.5)

dividing the mass percentage compositions by their respective molar masses. The larger number of moles is then divided by the smaller, to obtain the ratio needed for writing the simple empirical formula. Do not expect the calculated ratio to lead to an exact integer. Experimental measurements are subject to errors and uncertainty. If the combustion is incomplete or product lost during the experiment, the empirical formula that is determined will by oxygen deficient. If for some reason there are still drops of water (water is added to decompose any side products formed) in the crucible at the time of the final mass determination, then the recorded mass of product will be too large, and the simple empirical formula will have too much oxygen.

A number of ionic compounds contain one or more waters of hydration in their chemical formulas. A good example of this is potassium aluminum sulfate dodecahydrate (referred to as alum – $KAl(SO_4)_2 \cdot 12H_2O_{(solid)}$) which exists as a dodecahydrate. Alum will be synthesized in an experiment performed later this semester. As part of today's laboratory experiment you are going to study a different hydrate – $CuSO_4 \cdot xH_2O_{(solid)}$. The number of water molecules present in each formula of $CuSO_{4(solid)}$ can be determined by simply heating a known mass of $CuSO_4 \cdot xH_2O_{(solid)}$ over a Bunsen burner for a few minutes to drive off the water. The amount of water in the sample is

mass of water = mass of solid before heating – mass of solid after heating (3A.6)

determined by difference. The mass of the solid after heating is the mass of the anhydrous $CuSO_{4(solid)}$. The relative numbers of moles of $CuSO_4$ and H_2O in the hydrated sample are calculated as:

moles of $CuSO_4$ = mass of $CuSO_4$/molar mass of $CuSO_4$ (3A.7)

moles of H_2O = mass of H_2O/molar mass of H_2O (3A.8)

by dividing the mass of $CuSO_4$ and H_2O by their respective molar masses. The numerical of value of "x" is

$$x = \text{moles of } H_2O/\text{moles of } CuSO_4 \qquad\qquad (3A.9)$$

Do not expect the calculated ratio to lead to an exact integer. Remember experimental measurements are subject to errors and uncertainty. [Note – $MgSO_4 \cdot xH_2O_{(solid)}$ is another solid that could be used should one decide to study more than one hydrate.]

COMBUSTION OF MAGNESIUM STRIP

The metal (or porcelain) crucible and lid are cleaned by setting them on a ceramic triangle, and heating strongly with a Bunsen burner for approximately 5 minutes. When finished, only handle the crucible and lid with crucible tongs. Fingerprints may adversely affect the mass determinations. Allow the crucible and lid to cool to ambient room temperature. Weigh and record the mass of the crucible + lid on the Data Sheet under the column labeled Trial 1.

Next cut 15 cm of magnesium ribbon. Before you actually cut the magnesium ribbon into small strips make sure that it is clean and shiny. If the strip shows signs of corrosion (oxidation – dull gray or black coating), gently rub the magnesium with the steel wool, sandpaper or emery board that is provided. Now cut the strips into small pieces. Place the strips into the crucible, put the lid on the crucible and reweigh. Record the mass in the appropriate space on the Data Sheet. The smaller strips will lead to more efficient combustion.

Place the crucible with lid on the ceramic triangle, and heat gently with a Bunsen burner. Occasionally lift the lid to give the reaction oxygen. You want a controlled combustion so that no product escapes. If the magnesium were to be completely exposed to air (*e.g.*, no lid), then the metal would burn with a bright flame and part of the magnesium would likely be lost due to vaporization. When it appears that all of the magnesium has reacted, remove the lid, and continue heating for another minute or two. Now remove the Bunsen burner, and very carefully add a few drops of deionized water to decompose any side products. **DO NOT ADD MORE THAN A FEW DROPS OF WATER. TOO MUCH WATER WILL CAUSE THE CRUCIBLE TO COOL TOO FAST, AND THE CRUCIBLE WILL CRACK**. After the few drops of water are added, resume heating for another thirty seconds.

Now remove the heat, and allow the crucible to come to ambient room temperature. Weigh the crucible, lid and contents. Be sure to use the tongs to handle the crucible and lid. Write the mass down somewhere on a piece of paper. Do not record the mass yet on the Data Sheet as the combustion process may not be complete.

To determine whether all of the magnesium has reacted, heat the crucible for another minute. Let the crucible cool to room temperature and reweigh. If the new mass is within $\pm 5\%$ of the mass obtained when you weighed the product above, then the combustion is complete. Now record the mass of the crucible + lid + product on the Data Sheet. If not, part of the magnesium was still unreacted. Put the crucible back on the ceramic triangle and reheat for one more minute. Cool and reweigh. Repeat this step until a constant mass (± 5 %) is obtained.

You are now finished with Trial 1. The entire experiment needs to be performed a second time. Record the experimental values for the second trial on the laboratory Data Sheet under the column labeled Trial 2.

Calculate the Mg:O mole ratio by substituting the measured values into Eqns. 3A.1 – 3A.5. Calculate the average Mg:O mole ratio, and give the value to the TA who will then compile the values for the entire class. The data for the entire class will be posted on the bulletin board outside the laboratory room sometime during the next day. Go to the bulletin board outside the laboratory room and copy down the class results on your laboratory Data Sheet in the space provided. Calculate the class mean (average) and standard deviation for the Mg:O mole ratio.

DETERMINATION OF NUMBER OF WATERS OF HYDRATION

Clean the crucible and lid with tap water and then rinse with deionized water. Place the crucible with cover on the ceramic triangle. Make sure the lid is slightly askew so that water can escape when the crucible is heated (see Figure 3.1). Gently heat the crucible and lid over a Bunsen burner for about five minutes. Once the crucible has cooled to room temperature, weigh the crucible with lid on the top loading balance. Be sure to handle the crucible and lid with tongs as you did in the magnesium-oxide determination. Record the mass of the crucible with lid on the Data Sheet in the section labeled "Water of Hydration Determination." Now place about 1.000 to 1.500 grams of $CuSO_4 \cdot xH_2O_{(solid)}$ in the crucible and reweigh. Be sure to record the mass on the Data Sheet. Place the crucible with sample and lid (slightly askew so water can escape) on the ceramic triangle and heat for ten minutes. This should be sufficient time to remove all of the water molecules. Let the crucible to cool. Once the crucible has cooled to room temperature re-determine the mass of the crucible + contents + lid. Write the mass down somewhere on a piece of paper. Do not record the mass yet on the Data Sheet as the compound may still have some of its waters of hydration. To determine whether all of the water has been removed, heat the crucible for another minute. Let the crucible cool to room temperature and reweigh. If the new mass is within ±5 % of the mass obtained when you weighed the product above, then all of the water has been removed. Now record the mass of the crucible + lid + anhydrous copper sulfate on the Data Sheet. If not, put the crucible back on the ceramic triangle and reheat for two additional minutes.

Cool and reweigh. Repeat this step until a constant mass (± 5 %) is obtained. If time permits repeat the water of hydration determination a second time.

Calculate the number of waters of hydration by substituting the measured values into Eqns. 3A.6 – 3A.9.

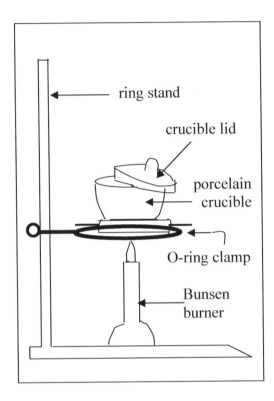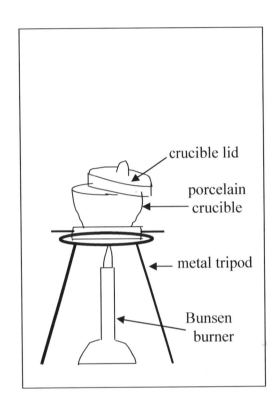

Figure 3A.1. Apparatus for the heading of a solid for the determination of the simple empirical formula of Mg_xO_y, and for the number of waters of hydration of $CuSO_4 \cdot xH_2O$. Either apparatus can be used in Experiment 3. The metal (or porcelain) crucible sits on a ceramic clay triangle, which is supported by either an O-ring clamp (left-hand drawing) or a metal tripod (right-hand drawing).

NOTE TO INSTRUCTOR – The laboratory manual was written with the intention that students would perform either Experiment 3A or Experiment 3B. The two experiments involve the determination of the simple empirical formula of a metal oxide. Experiment 3A involves the combustion of the metal oxide; whereas Experiment 3B involves the reduction of the metal cation back to the metallic element.

DATA SHEET – EXPERIMENT 3A

Name: _____

Simple Empirical Formula Determination for Magnesium Oxide

	Trial 1	Trial 2
1. Mass of crucible and lid, g:	_____	_____
2. Mass of magnesium at start, g:	_____	_____
3. Mass of crucible, lid and product, g:	_____	_____
4. Mass of product, g:	_____	_____
5. Mass of oxygen in product, g:	_____	_____
6. Moles of magnesium in product:	_____	_____
7. Moles of oxygen in product:	_____	_____
8. Mole ratio, Mg:O :	_____	_____
9. Average mole ratio, Mg:O :	_____	
10. Probable simple empirical formula:	_____	

Waters of Hydration Determination

	Trial 1	Trial 2
11. Mass of crucible and lid, g:	_____	_____
12. Mass of crucible, lid & $CuSO_4 \cdot xH_2O$ start, g:	_____	_____
13. Mass of $CuSO_4 \cdot xH_2O$	_____	_____
13. Mass of crucible, lid and $CuSO_4$ end, g:	_____	_____
14. Mass of water lost, g:	_____	_____
15. Mass of $CuSO_4$, g:	_____	_____
16. Moles of water:	_____	_____
17. Moles of $CuSO_4$:	_____	_____
18. Number of moles of water of hydration:	_____	

Class Data for Mg:O Mole Ratios

_____	_____	_____	_____
_____	_____	_____	_____
_____	_____	_____	_____
_____	_____	_____	_____
_____	_____	_____	_____
_____	_____	_____	_____

Class mean: _____

Class standard deviation: _____

EXPERIMENT 3B: DETERMINATION OF SIMPLE EMPIRICAL FORMULA – COMPARISON OF RED AND BLACK COPPER OXIDE FORMULAS

INTRODUCTION

Whenever a new compound is made, one of the first items of interest is its formula. That is how many atoms of each element is present in one molecule of the newly synthesized compound. This information is often determined by taking a weighed sample of the compound and either decomposing it into its component elements or reacting the compound with oxygen to produce substances such as carbon dioxide, water and nitrogen gas, which are then collected and quantified. The experimental results from such analyses provide the mass percentage of each type of element present in the compound. The mass percentages are then used to deduce the compound's simple empirical formula.

The simple empirical formula gives the relative number of atoms of each element present in the molecule. The molecular formula is more informative, in that this formula provides the actual numbers and types of atoms in the molecule. There are several compounds that have a simple empirical formula of CH. Acetylene (C_2H_2), benzene (C_6H_6), cyclobutadiene (C_4H_4) and styrene (C_8H_8) are four such molecules. Knowledge of the compound's molar mass enables one to get from the simple empirical formula to the molecular formula. One would need the structural formula (also called the molecular structure) in order to know how the atoms are actually attached (bonded) in the molecule. Of the three formulas, the simple empirical formula is the least informative; however, it is the easiest one to determine. The structural formula is the most informative, and as one might expect, it is the hardest of the three to determine. This semester you will learn experimental methods for determining both the simple empirical formula and the molecular formula. Next year in the organic chemistry courses you will learn several spectroscopic methods that can be used in molecular structure determinations.

In today's laboratory experiment you will determine the simple empirical formula of a compound containing only copper and oxygen. The copper oxide compound can be purchased commercially, either as a red or as a black solid material. Since the colors are different, the natural question that might be asked is "Are the simple empirical formulas also different?" Intuitively, one might expect the formulas to be different. There are some substances though which have different colors, yet the simple empirical formula is the same. Elemental phosphorous is one such example.

The simple empirical formula, Cu_xO_y, of the red and black forms of copper oxide will be determined by dissolving the solid in hydrochloric acid

$$Cu_xO_{y(solid)} + 2y\ HCl_{(aq)} \longrightarrow x\ Cu^{(2y/x)+}_{(aq)} + y\ H_2O + 2y\ Cl^-_{(aq)}$$

and then reacting the dissolved copper ion with aluminum foil

$$3\ Cu + (2y/x)\ Al_{(solid)} \longrightarrow (2y/x)\ Al^{3+}_{(aq)} + 3\ Cu_{(solid)}$$

to form metallic copper. The subscripts "x" and "y" in the formula denote the relative number of atoms of Cu and O in the simple empirical formula.

The number of atoms is proportional to the number of molecules, with the proportionality constant being Avogadro's number. It is more convenient to work in mole numbers as opposed to atom numbers. By knowing the mass of the copper oxide that was dissolved, and the mass of copper metal formed, one is able to calculate the number of moles of copper

number of moles of copper = mass of copper formed/molar mass of copper

(3B.1)

and number of moles of oxygen

number of moles of oxygen = mass of oxygen/molar mass of oxygen   (3B.2)

$$moles\ of\ oxygen = \frac{mass\ of\ copper\ oxide\ dissolved - mass\ of\ copper\ metal\ formed}{molar\ mass\ of\ oxygen}$$

(3B.3)

present in the dissolved copper oxide salt. The larger number of moles is then divided by the smaller value, to obtain the ratio needed for writing the simple empirical formula. Do not expect the calculated ratio to lead to an exact integer. Experimental measurements are subject to errors and uncertainty.

REACTION WITH ALUMINUM FOIL

Accurately weight about 1.0 grams of red copper oxide on an electronic top loading balance. Record the mass of the copper oxide to 3 decimal places on your Laboratory Data Sheet in the appropriate space under the section heading of "Empirical Formula of Red Copper Oxide." Place the copper oxide in a clean 400-ml beaker, and add 30 ml of 3 Molar hydrochloric acid (HCl) with a graduated cylinder. A slightly oversized beaker is being used to reduce loss of material due to splattering.

CAUTION: AVOID SPILLING THE HYDROCHLORIC ACID ON YOUR SKIN. AVOID BREATHING THE FUMES OR GETTING THEM INTO YOUR EYES.

Stir the resulting mixture with a glass stirring rod to facilitate the dissolution of the copper oxide. **Do not use a metal spatula as it will react with the solution**. The solution can be heated for 1 to 2 minutes to help dissolve part of the solid. The solution will likely turn bluish-green. After the solution has sat for 5 minutes, add several strips of food-grade aluminum foil to the solution (about 0.25 grams or less) to react with the copper ions. (**Note to Instructors** – Students should be cautioned to add only 2-3 pieces of aluminum at a time, and then to wait until the aluminum has reacted before adding any more strips. Students who add a large excess of aluminum will have to wait a long time for the excess to react with the acid.) Aluminum should be added until the bluish-green color disappears. When the copper oxide has completely reacted with the aluminum, the only solid remaining in the beaker is copper metal and some excess aluminum metal (no red/white or black solids). At this point in time the solution is likely to be cloudy from the hydrogen bubbles being formed at the surface of the aluminum foil, and the color may appear dark gray in color. Once the aluminum has finished reacting, there should be no more bubbles from the solid, and the solution should appear clear, or perhaps slightly yellow in color. Additional hydrochloric acid, HCl, can be added to react with any excess aluminum

$$6 \, H^+_{(aq)} + 2 \, Al_{(solid)} \, ---> 2 \, Al^{3+}_{(aq)} + 3 \, H_{2(gas)}$$

It is imperative that all excess aluminum foil be reacted as the solid residue is to be collected and weighed. The mass of unreacted aluminum foil will adversely affect the calculations. The reaction of hydrogen ions and aluminum metal is an oxidation-reduction reaction, where electrons are transferred from the aluminum to the hydrogen ion.

When all of the aluminum metal is gone, it is time to collect the copper metal. You will need to weigh a circle of filter paper and a clean watch glass, and record the mass of each in the appropriate space on your Data Sheet. Set up a glass funnel using a ring stand and iron ring or funnel support. Place the stem of the funnel in a 150-ml beaker. See Figure 3B.1 for filtration apparatus setup. Fold the filter paper into quarters (see Figure 3B.2), and place it in the funnel. Pour a little deionized water into the filter so that it will stay in place. Discard any water that runs through the filter into the beaker. Collect the copper metal by pouring the reaction mixture through the filter paper. The metallic copper will be retained on the filter paper. Be sure to transfer all solid from the beaker to the filter paper. You do not want any copper metal to remain behind as this will adversely affect the experimental results. After the liquid has drained, rinse the copper metal with a small amount of deionzied water, and then with a small of acetone. Carefully remove the filter paper form the funnel, and place on the watch glass that you previously weighed. Try not to lose any solid when removing the filter paper. Spread the filter paper out on the watch glass to dry. Once the filter paper and solid are dry, reweigh the watch glass plus filter paper plus copper metal. Calculate the weight of the copper by difference. The weight of the watch glass and filter paper are both known.

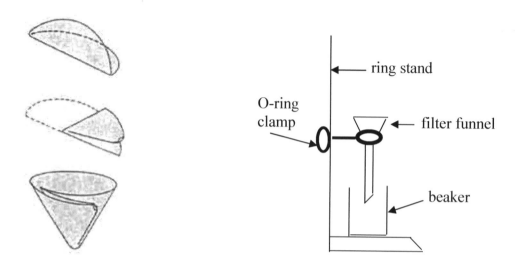

Figure 3B.2. Folding of a circle of filter paper

Figure 3B.1 Setup of filtration assembly Folded filter paper is placed into the filter funnel.

While you are waiting for the filter paper and copper metal to dry, repeat the entire experimental procedure using a sample of black copper oxide. Record the values for this part of the experiment on your Data Sheet under the section heading of "Simple Empirical Formula of Black Copper Oxide." Once all of the experimental data has been collected, determine the empirical formulas of both compounds by substituting the respective masses of the copper oxide solid and copper metal formed into Eqns. 3B.1 and 3B.3 Do red copper oxide and black copper oxide have the same simple empirical formula?

Give the value of the Cu:O mole ratio for each compound to the TA who will then compile the values for the entire class. The data for the entire class will be posted on the bulletin board outside the laboratory room sometime during the next day. Go to the bulletin board outside the laboratory room and copy down the class results on your laboratory Data Sheet in the space provided. Calculate the class mean (average) and standard deviation for the Cu:O mole ratios for both the red and black copper oxides.

NOTE TO INSTRUCTOR – The laboratory manual was written with the intention that students would perform either Experiment 3A or Experiment 3B.

DATA SHEET – EXPERIMENT 3B

copper oxide

Name: _____

Simple Empirical Formula of Red Copper Oxide:

1. Mass of red copper oxide solid, g: _____

2. Mass of filter paper, g: _____

3. Mass of watch glass, g: _____

4. Mass of watch glass + filter paper + copper metal, g: _____

5. Mass of copper metal formed, g: _____

6. Mass of oxygen in product, g: _____

7. Moles of copper in red copper oxide sample: _____

8. Moles of oxygen in the red copper oxide sample: _____

9. Mole ratio, Cu:O :   _____

10. Probable simple empirical formula: _____

Simple Empirical Formula of Black Copper Oxide:

11. Mass of red copper oxide solid, g: _____

12. Mass of filter paper, g: _____

13. Mass of watch glass, g: _____

14. Mass of watch glass + filter paper + copper metal, g: _____

15. Mass of copper metal formed, g: _____

16. Mass of oxygen in product, g: _____

17. Moles of copper in red copper oxide sample: _____

18. Moles of oxygen in the red copper oxide sample: _____

19. Mole ratio, Cu:O :   _____

20. Probable simple empirical formula: _____

Comparison of Simple Emprical Formulas

21. Are the simple empirical formulas of red copper oxide and
    black copper oxide the same (yes or no)?   _____

Class Data for Cu:O Mole Ratios for Red Copper Oxide

_____	_____	_____	_____
_____	_____	_____	_____
_____	_____	_____	_____
_____	_____	_____	_____
_____	_____	_____	_____
_____	_____	_____	_____

Class mean: _____

Class standard deviation: _____

Class Data for Cu:O Mole Ratios for Black Copper Oxide

_____	_____	_____	_____
_____	_____	_____	_____
_____	_____	_____	_____
_____	_____	_____	_____
_____	_____	_____	_____
_____	_____	_____	_____

Class mean: _____

Class standard deviation: _____

EXPERIMENT 4: DETERMINATION OF THE MOLAR MASS OF A VOLATILE LIQUID

INTRODUCTION

The percent elemental composition of a compound, such as was determined in Experiment 3, enables one to write the simple empirical formula. The transition from the empirical formula to the molecular formula requires one additional piece of information, namely the molar mass. For a molecular compound having the empirical formula of CH, there is an infinite number of molecular formulas (C_2H_2, C_4H_4, C_6H_6, C_8H_8, $C_{10}H_{10}$, ..., $C_{2n}H_{2n}$) that correspond to a carbon : hydrogen ratio of 1:1. If the molar mass of the compound is known, then one can narrow down the list of possibilities to only those few compounds possessing the specified molar mass. In the example above, if the molar mass of the compound is known to be 78.11 grams/mole, then the molar molecular formula must be C_6H_6. This is the only molecular formula amongst the list of possibilities that has the correct molar mass.

During the two-semester general chemistry laboratory sequence you will perform several different experiments in which the molar mass of a compound is determined. Some of the experimental methods will be limited in application to volatile solutes, whereas other methods will be better suited for nonvolatile compounds. The first of the molar mass determinations will be performed today, and is based on the ideal gas law. The ideal gas law states that

$$\text{Pressure} \times \text{volume} = \text{moles of gas} \times \text{gas constant} \times \text{temperature} \qquad (4.1)$$

the pressure-volume product of a gas equals the number of moles of the trapped gas times the ideal gas law constant times the absolute Kelvin temperature. Expressed in terms of the molar mass, Eqn. 4.1 becomes

$$\text{Pressure} \times \text{volume} = (\text{grams of gas/molar mass of gas}) \times R \times \text{Temperature}$$

$$(4.2)$$

In Equations 4.1 and 4.2 the pressure is given in atmospheres, the volume in liters and the temperature in Kelvin. The ideal gas law constant equals R = 0.08206 liter-atm/(mole Kelvin). Later this semester you will perform an experiment in which the ideal gas law is derived. For now though, the ideal gas law will be used strictly as a mathematical equation. Equation 4.2 contains 6 variables. If any five of the values are known, the remaining sixth variable can be calculated. The molar mass is calculated by trapping a gas sample at a known temperature and pressure, and then measuring the mass and volume of the trapped gas. All of the quantities in Eqn. 4.2 are known, except for the molar mass of the gas that you want to determine.

MOLAR MASS DETERMINATION BASED ON WATER DISPLACEMENT METHOD

The molar mass of butane is to be determined using a water displacement method. Weigh a cigarette lighter and record the mass on the laboratory Data Sheet in the appropriate place in the section labeled "Molar Mass of Butane Determination". Fill a plastic trough full of water, and record the temperature of the water. Make sure that the water is at the same temperature as the laboratory room. Record the value on the Data Sheet. You will need to convert this value to Kelvin for the molar mass calculation. Next dunk a 50-ml graduated cylinder in the water and invert. Make sure that no gas has been trapped in the cylinder. Now carefully place the cigarette lighter under the inverted graduated cylinder and release a quantity of butane gas into the cylinder. You may need to tip the graduated cylinder some. When the gas has been collected take the volume reading. It is important for the water level inside the graduated cylinder and outside the cylinder be as close to each other as possible. Otherwise the pressure of the trapped gas will not be equal to the atmospheric pressure. One would need to correct for difference in the height of the two water levels.

Carefully remove the cigarette lighter, blot with a paper towel, air dry and weigh. The mass of the lighter must be less than at the start of the experiment. If the mass is heavier, then you have not thoroughly dried the lighter. Record the mass of the lighter after butane release on the Data Sheet. The mass of the butane trapped is calculated by difference. Ask the TA for the current atmospheric pressure. The atmospheric pressure will likely be given to you in mm-Hg – you need to convert to atm by dividing the value by 760 mm-Hg/atm. Repeat the experimental procedure a total of five times.

Calculate the molar mass of butane by substituting the measured values into Eqn. 4.2. Carefully examine the calculated values to determine whether or not any of the values can be discarded (perform the Q-test; discussed later in this Experiment). Calculate the mean (after discarding any values) and report the value to the TA who will then compile the values for the entire class. The data for the entire class will be posted on the bulletin board outside the laboratory room sometime during the next day. Go to the bulletin board outside the laboratory room and copy down the class results on your laboratory Data Sheet in the space provided. Calculate the class mean (average) and standard deviation for molar mass determination.

MOLAR MASS DETERMINATION BASED ON SYRINGE METHOD

A 60-ml plastic syringe (Luer-lok syringe with latex syringe cap) has been modified for performing this part of the experiment. On the plunger small holes (nail stops) has been drilled at a 45° angle to the "plastic rib". The holes are drilled to correspond to syringe volumes of 30 ml, 40 ml and 50 ml. The diameter of the holes should be such that a 2-inch finishing nail snuggly fits into the hole.

Determine the mass of the syringe at the nail stop by placing the latex cap on the syringe, and pulling the plunger back just past the nail stop. Put the finishing nail into the last hole (50 ml). The plunger should now be allowed to move slowly forward, until it is stopped by the nail that was just inserted. The plunger may be hard to pull back as a "vacuum" is being formed in the syringe. Have your lab partner help you. Now place the syringe on the balance in an upright position (standing on the handle of the plunger) and weigh. Record the mass of the evacuated syringe at the nail stop on the laboratory Data Sheet.

It is now time to trap and weigh the gas samples. Large molar mass gases work best for this experiment. Air and carbon dioxide are two possibilities. If air is used, remove the syringe cap, pull the plunger back just passed the first nail stop, insert the finishing nail, and gently push the plunger move forward until it is stopped by the inserted nail. Put the latex cap on the syringe. Reweigh the syringe and trapped gas. Record the mass and volume of trapped gas on the Data Sheet. For this experiment, the volume of gas is measured using the markings on the syringe. [If greater accuracy is desired, one can fill the syringe with water, and record the mass of water contained in the syringe at each nail stop. The volume is calculated from the density of water, *e.g.*, each gram of water corresponds to 1 ml. Wait until the end of the experiment if you decide to determine the volume in this fashion. It will take awhile for the syringe to dry thoroughly, and you still have one more trapped gas measurement to make.]

If you decide to trap carbon dioxide gas, rather than air, place the tip of the syringe close to the evaporating dry ice. Fill the syringe and expel the gas a couple of times to remove any trapped air. Now trap the gas sample by pulling the plunger back past the nail stop, insert the nail stop, and gently push the plunger forward until it is stopped by the nail. The trapped gas sample will likely be colder than room temperature. Leave the latex cap off until the barrel feels like it is at room temperature. Now put the cap on the syringe, and weigh. Record the mass and volume of trapped gas on the Data Sheet.

Perform the experiment two times. Once the experimental measurements are over calculate the molar mass of the gas by substituting the measured values into Eqn. 4.2. Of the two experimental methods, which one do you think is the best?

REJECTION OF DATA – Q-test

As the experiments become more complex, there is a greater likelihood that you will make a mistake in following the written experimental procedure. Perhaps you heated the solution to too high of an experimental temperature to cause undesired chemical decomposition. Perhaps you added the wrong chemicals to the solution. Perhaps you missed the endpoint of a titration by adding reactant well beyond the desired color change. In such cases the experimental may give a set of measured values that appears to be skewed by the presence of one or more questionable data points that are not such consistent with the remaining data points. Such values are sometimes referred to as outliers. Inclusion of questionable values in the statistical data treatment may have a significant effect on the calculated mean and standard deviation.

Under what circumstances can experimental data be discarded, and not included in the data treatment? One justification for rejecting experimental data is quite obvious – we know with absolute certainty that a mistake was made. If the experimental procedure instructed me to stop the addition of sodium hydroxide when the solution turns faint pink, and I kept adding sodium hydroxide well beyond the noted color change, should I be surprised if the resulting experimental value was far different from the replicate determinations when the instructions were carefully followed? I should have started the experiment over when I first noticed that a procedural error had been made.

The second justification for rejecting experimental values is based more on a statistical argument. From the earlier discussion of random errors, you know that smaller random errors are more likely than larger random errors. Moreover, random errors follow a Gaussian distribution where about 95 % of the area under the curve falls between ± 3 standard deviations of the mean. If an outlier data point falls more than 3 standard deviations from the mean value, it is very likely that the large distance from the mean was not the result of a random error. If this is the case, then one must conclude that a non-random error had been made, in other words, a systematic error (*e.g.*, mistake) had been made.

The Q-test (referred to as Dixon's test, or Dixon's Q method in some books) is based along a similar line of reasoning. To apply the Q-test, the experimental data are arranged in either ascending or descending order. The *gap* is defined as the distance between the questionable value and its nearest neighbor. The *range* is the total spread of the data, from the lowest value to the highest value. A value of Q_{calc} is

$$Q_{calc} = gap/range \tag{4.3}$$

calculated as the ratio of the gap divided by the range. If $Q_{calc} > Q_{table}$ for whatever level of confidence that is being tested at, then the questionable value can be rejected. On the other hand, if $Q_{calc} < Q_{table}$ then the value must be retained. Table 4.2 lists Q-values for various levels of confidence and number of experimental observations. You will notice that it its progressively harder to reject a second experimental value, once the first value is discarded. The value of Q_{table} increases as more data points are removed.

Table 4.2. Values of Q for Rejection of Data*

Observations	Confidence Level		
	90 %	95 %	99 %
3	0.94	0.97	0.99
4	0.76	0.83	0.93
5	0.64	0.71	0.82
6	0.56	0.62	0.74
7	0.51	0.57	0.68
8	0.47	0.53	0.63
9	0.44	0.49	0.60
10	0.41	0.47	0.57
15	0.34	0.38	0.48
20	0.30	0.34	0.42
25	0.28	0.32	0.39

*A more extensive tabulation of Q-values can be found in D. B. Rorabacher, *Anal. Chem.*, **63**, 139-149 (1991); W. J. Dixon, *Ann. Math. Stat.*, **22**, 68-78 (1951); and *Statistics for Analytical Chemists*, R. Caulcutt and R. Boddy, Chapman and Hall, New York, NY (1983).

WATER DISPLACEMENT METHOD – CORRECTION TO TAKE INTO ACCOUNT THE VAPOR PRESSURE OF WATER

Now that the experiment is over, think very carefully about the experimental procedure in the molar mass determination based on the water displacement method. There is a flaw in the calculation procedure. Do you know what it is? The trapped gas in the graduated cylinder is not all butane. A small bit of water has evaporated into the space, and one needs to make a correction for the pressure that the evaporated water exerts on the solution. The pressure of the total gas inside the graduated cylinder is

pressure of trapped gas = partial pressure of butane + partial pressure of water

(4.4)

which should be

pressure of atmosphere = partial pressure of butane + partial pressure of water

(4.5)

equal to atmospheric pressure (assuming that the water two water levels were equal). How big of a difference does this make in the molar mass of butane determination? Calculate the partial pressure of butane in the trapped gas sample by subtracting the partial pressure of water (see Table 4.3 for water vapor pressure data) from the atmospheric pressure. Tabulated values should be adequate for the ambient room temperatures typically encountered in the general chemistry laboratory. More extensive tabulations can be found elsewhere (for example, see *Lange's Handbook of Chemistry*, 13th Edition, J. A. Dean, McGraw-Hill, New York, NY (1985); *CRC Handbook of Chemistry and Physics*, 71st Edition, D. R. Linde, CRC Press, Boca Raton, FL (1990); htpp://webbook.nist.gov). Now calculate the molar mass of butane using the corrected partial pressure of butane in Eqn. 4.2. How big of a difference did not taking the water vapor into account in the molar mass determination make?

Table 4.3. Vapor Pressure of Water at Several Temperatures

Temperature (°C)	Pressure (mm Hg)
17	14.5
18	15.5
19	16.5
20	17.5
21	18.6
22	19.8
23	21.1
24	22.4
25	23.8
26	25.2
27	26.7
28	28.3
29	30.0
30	31.8

DATA SHEET – EXPERIMENT 4

Name: _____

Molar Mass Determination of Butane Based on Water Displacement Method

	Trial 1	Trial 2	Trial 3
1. Lighter mass before gas release, g:	_____	_____	_____
2. Lighter mass after gas release, g:	_____	_____	_____
3. Mass of released gas, g:	_____	_____	_____
4. Volume of gas released, ml:	_____	_____	_____
5. Volume of gas released, l:	_____	_____	_____
6. Pressure of atmosphere, mm Hg:	_____	_____	_____
7. Pressure of atmosphere, atm:	_____	_____	_____
8. Temperature of gas, °C:	_____	_____	_____
9. Molar mass of gas, grams/mole:	_____	_____	_____

	Trial 4	Trial 5
1. Lighter mass before gas release, g:	_____	_____
2. Lighter mass after gas release, g:	_____	_____
3. Mass of released gas, g:	_____	_____
4. Volume of gas released, ml:	_____	_____
5. Volume of gas released, l:	_____	_____
6. Pressure of atmosphere, mm Hg:	_____	_____
7. Pressure of atmosphere, atm:	_____	_____
8. Temperature of gas, °C:	_____	_____
9. Molar mass of gas, grams/mole:	_____	_____

10. Can any experimental values be discarded based on the Q-test: _____

11. Average molar mass (discarded values removed), grams/mole: _____

12. Average molar mass corrected for partial pressure of water: _____

Molar Mass Determination Based on Syringe Method

	Trial 1	Trial 2
12. Mass of evacuated syringe, g:	_____	_____
13. Mass of syringe + trapped gas, g:	_____	_____
14. Mass of trapped gas, g:	_____	_____
15. Temperature of gas, ºC:	_____	_____
16. Temperature of gas, K:	_____	_____
17. Volume of gas, ml:	_____	_____
18. Volume of gas, l:	_____	_____
19. Pressure of gas (atmospheric pressure), mm Hg:	_____	_____
20. Pressure of gas, atm:	_____	_____
21. Molar mass of trapped gas, grams/mole:	_____	_____
22. Average molar mass of trapped gas, grams/mole:	_____	

Class Data – Molar Mass of Butane:

_____	_____	_____	_____
_____	_____	_____	_____
_____	_____	_____	_____
_____	_____	_____	_____
_____	_____	_____	_____
_____	_____	_____	_____

Class mean for molar mass of butane, grams/mole: _____
    (Before any values are rejected)

Class standard deviation for molar mass of butane: _____
    (Before any values are rejected)

Class mean for molar mass of butane, grams/mole: _____
    (After any values are rejected)

Class standard deviation for molar mass of butane: _____
    (After any values are rejected)

EXPERIMENT 5: PREPARATION OF ALUM

INTRODUCTION

Understanding chemical reactions is an important facet of our study of chemistry. It is not adequate to simply know that hydrogen gas reacts with nitrogen gas to form ammonia. Much more information is needed in order to produce ammonia on an industrial scale. Does one need to heat the two reactants in order to get them to react? If so, is there an optimum temperature at which a maximum amount of product is formed? How long does it take to produce a given amount of ammonia? Can one shorten the time by changing the temperature, or by starting with different amounts of hydrogen and/or nitrogen gas? If so, how much can the time be shortened without sacrificing the efficiency of the reaction? Speaking of reaction efficiency, what is the efficiency of the reaction, expressed as the amount of ammonia produced divided by the amount of starting materials? Can the efficiency be increased, and if so, how? These are just a few of the questions that one must address as laboratory bench-top chemical reactions are scaled up to a commercial manufacturing process.

In today's laboratory experiment, you will prepare alum ($KAl(SO_4)_2 \cdot 12H_2O$ – potassium aluminum sulfate dodecahydrate) by dissolving aluminum metal in a potassium hydroxide solution:

$$2\,Al_{(solid)} + 2\,KOH_{(aq)} + 6\,H_2O \longrightarrow 2\,KAl(OH)_{4(aq)} + 3\,H_{2(gas)}$$

followed by the addition of sulfuric acid

$$2\,KAl(OH)_{4(aq)} + H_2SO_{4(aq)} \longrightarrow 2\,Al(OH)_{3(solid)} + K_2SO_{4(aq)} + 2\,H_2O$$

$$2\,Al(OH)_{3(solid)} + 3\,H_2SO_{4(aq)} \longrightarrow Al_2(SO_4)_{3(aq)} + 6\,H_2O$$

The first chemical reaction is an oxidation-reduction reaction; the latter two reactions are acid-base reactions. If only a small amount of H_2SO_4 is added, the reaction stops at $Al(OH)_{3(solid)}$. If more sulfuric acid is added, the reaction goes all the way to $Al_2(SO_4)_{3(aq)}$, which contains K^+, Al^{3+} and SO_4^{2-} ions in the proportions in which they are found in alum. After sufficient cooling

$$K^+_{(aq)} + Al^{3+}_{(aq)} + 2\,SO_4^{2-}_{(aq)} \longrightarrow KAl(SO_4)_2 \cdot 12H_2O_{(solid)}$$

crystals of hydrated potassium aluminum sulfate, or alum, will form. Alum is particularly disposed to crystallization. It is used in dyeing fabrics, tanning leather and clarifying water in municipal water treatment facilities. Medicinally, alum has applications as an astringent and a styptic. For example, if a person knicks himself/herself while shaving, he/she might use a stick of alum to coagulate the blood and stop the bleeding. Alum is also found in several acme medications.

PREPARATION OF ALUM

Cut a piece of aluminum metal into small pieces and weight out about 1 gram on a top loading electronic balance. Record the mass of the aluminum to the third decimal on the laboratory Data Sheet. The aluminum metal is the limiting reagent for today's laboratory experiment, and the mass of Al must be known in order to calculate the maximum expected yield (*e.g.*, theoretical yield). Next place the aluminum in a 400-ml beaker, and add 50 ml of water (graduated cylinder) and 10 ml of 6 Molar KOH (graduated cylinder). Place the beaker on a wire gauze on a ring stand and gently heat.

CAUTION – H_2 GAS (VERY EXPLOSIVE) IS PRODUCED. BE SURE THAT YOU ONLY HEAT THE BOTTOM OF THE BEAKER AND DO NOT LET THE FLAMES GET NEAR THE TOP.

When the bubbles have stopped, remove the heat. Filter any solid residue remaining, and collect the clear filtrate in a 100-ml beaker. Next slowly add 20 ml of 9 Molar H_2SO_4 (graduated cylinder) to the filtrate.

CAUTION – SULFURIC ACID IS A STRONG ACID AND IS CAUSTIC. IT CAN CAUSE BURNS IF LEFT IN CONTACT WITH THE SKIN. IF ANY SULFURIC ACID GETS ON YOUR SKIN IMMEDIATELY RINSE WITH PLENTY OF COOL WATER FOR SEVERAL MINUTES. INFORM THE TEACHING ASSISTANT IF YOU SPILL ANY SULFURIC ACID ON THE BENCHTOP.

A white powder of $Al(OH)_{3(solid)}$ should form. Heat gently while stirring until the solution becomes clear. Add 2 or 3 boiling chips, and boil the solution down to a total volume of about 45 ml. While the solution is heating, prepare an ice-water bath by half filling a 400-ml beaker with ice, and then add water to fill the beaker to the 300-ml mark. Once the solution volume is down to 45 ml, turn off the gas to the burner, and allow the beaker to cool to room temperature. After the beaker has cooled, place it in the ice-water bath. Clear crystals of alum should form. Continue cooling for another 30 minutes to allow all of the alum to precipitate.

Once the 30-minute cooling period is over, the solid is collected by vacuum filtration, and washed with 20-ml of chilled ethanol to remove any adsorbed impurities. The vacuum filtration apparatus is constructed by attaching one end of rubber tubing to the side-arm of a vacuum flask, and the other end to an aspirator on the water faucet. When the water is turned on, the water flow past the aspirator creates suction. Place the Büchner filter funnel into the mouth of the side-arm vacuum filter flask. (See Figure 5.1 on the next page.) Place the filter paper inside the funnel and make sure that all of the

60

holes are covered. Note the filter paper stays flat during this process. Transfer the contents from the beaker to the Büchner funnel. Continue to draw air through the alum for another 10 to 15 minutes to dry the crystals. Remove the boiling chips.

Weigh an empty 150-ml beaker, and record the mass on the laboratory Data Sheet. Turn off the source of the vacuum. Carefully transfer the dried alum crystals and filter paper circle to the weighted 150-ml beaker. Use a metal spatula to scrape off any crystals adhering to the filter paper. Reweigh the beaker containing the alum crystals, and record the mass on the Data Sheet. Calculate the mass of the alum produced by difference.

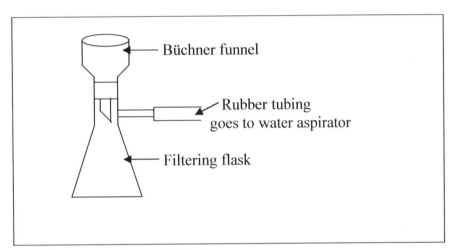

Figure 5.1 – Setup of filtering apparatus

PERCENT YIELD CALCULATION

Finally, calculate the percent yield:

Percent yield = 100 × (actual yield of alum/theoretical yield of alum)   (5.1)

The numerator in Eqn. 5.1 is the actual mass of alum that was experimentally made. The denominator is based on the limiting reagent. The molecular formula of alum, $KAl(SO_4)_2 \cdot 12H_2O_{(solid)}$, contains one Al atom. For every mole of aluminum metal used, one mole of alum should be produced:

moles of alum theoretical = moles of aluminum used   (5.2)

mass of alum expected/molar mass of alum = mass of aluminum/molar mass of aluminum   (5.3)

assuming that the reaction goes to completion, that is no product is lost, and that there are no side reactions that consume part of the aluminum. The molar mass of $KAl(SO_4)_2 \cdot 12H_2O = 474.08$ grams/mole. Substitute the mass of aluminum that was used into Eqn. 5.3, along with the molar masses of aluminum and alum, to calculate the theoretical yield of alum expected. Now calculate the percent yield by substituting the theoretical and actual yields into Eqn. 5.1.

The percent yield should fall between 0 % (no product formed) and 100 % (complete efficiency). Percent yield calculations exceeding 100 % violate the conservation of mass. What errors could possible cause one to calculate a percent yield that violates the conservation of mass? Well, in this particular experiment, the solid may not be completely dry or impurities may have become tramped in the solid during the crystallization. Both errors would cause the mass of the weighed solid to be too heavy. Percent yields less than 100 % are much more common. Few chemical reactions or experimental manipulations are 100 % efficient, despite carefully controlled laboratory conditions. Side reactions can occur that form products other than the desired one, and during the collection and purification steps, part of the desired product may be lost. Also as you will later discover, chemical reactions do not necessarily go 100 % to completion. Even though we sometimes write the chemical reaction with a unidirectional arrow, "------->", there may still be unreacted reactants. A partial reaction that you may have already discussed in the chemistry lecture course involves the dissociation of acids. Strong acids completely dissociate. Weak acids, on the other hand, only partially dissociate in water.

DATA SHEET – EXPERIMENT 5

Name: _____

Preparation of Alum:

1. Mass of aluminum pieces, g: _____

2. Mass of empty 150-ml beaker, g: _____

3. Mass of beaker + alum, g: _____

4. Mass of alum, g: _____

Calculations:

5. Moles of aluminum used: _____

6. Moles of alum expected: _____

7. Mass of alum expected, g: _____

8. Percent yield for alum preparation: _____

Crystal Shapes:

9. Alum is particularly disposed to crystallization and it is one of the chemicals included in many of the commercial crystal growing kits. Sketch the shape(s) of the alum crystal(s) that were collected in the Büchner funnel.

10. The seven basic crystal systems are illustrated below:

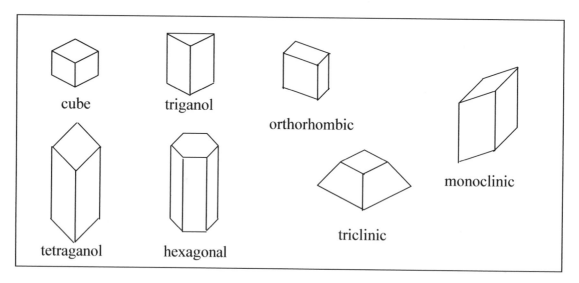

Structural features: (a) monoclinic – 2 of the 3 axes meet at 90 and all 3 axes have different lengths; (b) triclinic – all axes have different lengths and none meet at 90 ; (c) orthorhombic – all 3 axes meet at 90 and each axes has a different length; and (d) tetraganol – all 3 axes meet at 90 and 2 of the 3 axes have the same length. Additional information is given in Appendix E.

What basic shape(s) did you observe in the case of the alum crystals? (Note: Some of the alum may appear octahedron due to the rate of crystal growth on the sides versus crystal growth at the corners. Octahedron is a subclass within one of the seven basic crystal systems.)

EXPERIMENT 6: DENSITIES OF SOLIDS - IDENTIFICATION OF UNKNOWN MATERIALS AND DENSITY DETERMINATION THROUGH LINEAR REGRESSION ANALYSIS

INTRODUCTION

In today's laboratory experiment you will determine the density of several regular geometric-shaped and several irregular shaped objects of various sizes. The determination involves measuring both the object's mass and volume, which are then substituted into:

$$\text{density} = \text{mass of material/volume of material} \qquad (6.1)$$

the definition of density, Eqn. 6.1. Generally, in scientific applications the mass is expressed in grams and the volume in cm^3 (ml). For regular geometric-shaped items the volume can be determined through a mathematical formula:

Cube: $\qquad$ volume = (length of side)^3 $\qquad\qquad$ (6.2)

Sphere: $\qquad$ volume = (4/3) π (radius)^3 $\qquad\qquad$ (6.3)

Rectangular bar: $\qquad$ volume = length $\times$ width $\times$ height $\qquad$ (6.4)

Cylindrical rod: $\qquad$ volume = 4 π (radius)^2 $\times$ height $\qquad$ (6.5)

or by the method of liquid displacement. When completely submerged a solid object displaces a volume of liquid exactly equal to its own. Naturally, none of the solid material can dissolve, which places limitations on the chemical compounds that can be studied by the displacement method. Volumes of irregular shaped solids must be measured by liquid displacement. Densities of solid chemical substances can be used to help in compound identification, in much the same way that liquid densities were used in Experiment 2.

Today's laboratory experiment provides the opportunity for us to expand on our earlier discussion of experimental uncertainties. Not all of the density measurements have the same level of experimental uncertainty/error. This point can be illustrated with the following example. You will determine the density of several ball bearings made from the same material. Each ball bearing should then have the same density. For the

very large ball bearings, $\bullet$, the mass and volume can be determined with very little

relative experimental uncertainty. According to the manufacturer's specifications, the balance reading should be accurate to ±0.002 grams (for the three decimal-place electronic top loading balances). The recorded mass of a large ball bearing is: *e.g.*, mass = large numerical value ± 0.002 grams; so the uncertainty of ± 0.002 grams is quite negligible compared to the object's mass. Similarly, there will be an uncertainty associated with the volume measurement. For the ball bearings the volume will be determined by formula (volume of sphere = (4/3) π (radius)^3, after the diameter of the spherical ball bearing is measured with a manila-folder caliper. Rather than take the time to try to estimate the experimental uncertainty for this type of volume determination, let's just assume a guessed value of ± 0.10 cm^3. For a very large ball bearing, the experimental uncertainty is again quite small to the measured value, *e.g.*, volume = large numerical value ± 0.10 cm^3. How much confidence do you have in the density calculated from the measured mass and volume? Quite "confident" as there is very little relative uncertainty in the numerical values that were substituted into the numerator and denominator of Eqn. 6.1. You would expect the calculated density to be of "very good" quality.

Now lets move to the middle-sized ball bearings, •. The recorded mass is not as large as before, *e.g.*, mass = middle numerical value ± 0.002 grams; however, the

experimental uncertainty of ± 0.002 grams is still small compared to the value itself. The same is true for the volume, *e.g.*, volume = middle numerical value ± 0.10 cm^3. The density of the middle-sized ball bearing calculated from these values still should be of "fairly good" quality, but perhaps not as good as that of the larger-sized ball bearings. Remember the larger-sized ball bearings had smaller relative uncertainties in both its measured mass and volume.

Finally we get to the very small-sized ball bearings, ·. The recorded mass is now quite small, *e.g*, mass = small numerical value ± 0.002 grams, and the experimental uncertainty of ± 0.002 grams is becoming increasing more significant with decreasing object mass. The same is true in regards to the volume of the ball bearing. It is quite conceivable that the experimental uncertainty in volume may be nearly as large as the volume itself, *e.g.*, volume = 0.15 ± 0.10 cm. How confident would one be in the density calculated from these two numerical values. Not very confident. For lack of a better word, let's just say the calculated density of the smallest ball bearing would be of "very poor quality". How would one obtain the density of the material that was used to make the ball bearings? Although all the ball bearings were made from exactly the same material, taking the simple arithmetic average (average or mean value) does not seem right. The different-sized ball bearings did not contribute equally in accuracy to the final value of the density. The smallest ball bearings, with a very small mass and a very small volume, contributed the largest relative error. A simple arithmetic average treats all of the data points the same, in other words the "very poor quality" data would be treated the same as the "very good quality" data. Statistical methods have been developed to deal with just this type of problem. One method would be to calculate a weighted standard deviation, where each individual data point could be weighted according to the reciprocal of its relative uncertainty/error. The method that will be used in today's laboratory experiment is based on a linear least squares regression analysis.

LINEAR LEAST SQUARES REGRESSION ANALYSIS

The ball bearing data set consists of a set of (volume, mass)-ordered data points. The points when graphed with the volume along the x-axis and mass along the y-axis, should fall on a straight line. The slope of this line, slope = change in mass/change in volume, corresponds to the density of the material used in the fabrication of the ball bearings. To get the "best" straight line through the data, one needs to minimize

$$\text{minimization function} = \Sigma \, (y_i^{exp} - y_i^{line})^2 \tag{6.6}$$

the sum of the distances between each data point and the line. It can be shown that minimizing the squares of the deviations (rather than simply their magnitudes) corresponds to assuming that the set of y–values is the most probable set. In minimizing only the vertical distances one assumes that the uncertainties in the y-values are much greater than the uncertainties in the x-values. Normally when making experimental measurements, one measured property does have more uncertainty than the other. However, in this case one could make a fairly persuasive argument that volume has the larger uncertainty of the two properties being measured. What we will do for our data treatment is to first perform the regression analysis with the mass graphed along the y-axis and volume graphed along the x-axis, and then repeat the analysis with the analysis with the properties reversed. In the latter analysis, the density of the material should be the reciprocal of the calculated slope

The equation of a straight line is: $y_i^{line} = (\text{slope}) \times x_i + \text{intercept}$. Substituting this linear relationship into Eqn. 6.6, the following result is obtained:

$$\text{minimization function} = \Sigma \, (y_i^{exp} - (\text{slope}) \times x_i - \text{intercept})^2 \tag{6.7}$$

You may already know from calculus that minima (or maximum) in a mathematical function is found by taking the derivative of the function, and then setting the derivative equal to zero. Since there are two quantities to be calculated, the slope and the intercept, two derivatives are needed:

$$\frac{\partial\, minimization\ function}{\partial\, slope} = 0 \qquad\qquad (6.8)$$

$$\frac{\partial\, minimization\ function}{\partial\, intercept} = 0 \qquad\qquad (6.9)$$

In the first differentiation, Eqn. 6.8 (which is really a partial derivative), the intercept is held constant. In the differentiation with respect to the intercept, Eqn. 6.9, the slope is held constant. Equations 6.8 and 6.9 are solved simultaneously for the numerical values of the slope and intercept of the "best" straight line. Fortunately, Pasco's built-in software performs this computation for us. Technically, the line should be forced through origin, but this is not one of the built-in options. The software also gives the correlation coefficient, r, which is a measure of how close the line falls to the experimental data points. If the set of data points falls exactly on the straight line, the correlation coefficient is exactly r = 1 (or r = -1 for a negative slope). For perfect randomness, with absolutely no correlation whatsoever, the correlation coefficient would be r = 0.

If you perform the experiment carefully you should get a correlation coefficient fairly close to r = 1, except for perhaps a bad data point. If a data point looks "bad" you might try removing the point from the data set, and perform the regression analysis over again. Remember that in this experiment the different-sized ball bearings did not contribute equally in accuracy to the final value of the density. The data point for the smallest ball bearing is expected to have the largest relative experimental uncertainty. A mathematical consequence of the linear least squares method is that the largest data points are weighted more heavily in the minimization procedure. Does the method of linear least squares analysis seem appropriate for the ball bearing data set? Yes, the largest ball bearings should have the smallest relative experimental uncertainty, and the final answer should be based more heavily on the "better" data points. The linear least squares method gives this type of weighting.

DENSITY DETERMINATION OF SOLID OBJECTS BY LIQUID DISPLACEMENT

The experimental procedure of this part of the laboratory exercise is relatively straightforward. The mass of each object is obtained by weighing the item on a top loading electronic balance. Be sure that each item is dry before weighing. Someone in the class may have used the unknown previously, and it may still be wet. If so, dry with a paper towel as you do not want to extraneous water included in the compound's mass. Record the mass on the laboratory Data Sheet in the section labeled "Density Determination by Liquid Displacement". To determine the volume, fill a graduated cylinder about half-way full with distilled water. You will need to add sufficient water so that the object is completely submerged, but not too much water, otherwise the final volume reading will be above the lines on the graduated cylinder. Record the volume of the water in the graduated cylinder on the laboratory Data Sheet in the appropriate space. In reading the volume of water in the graduated cylinder, the reading is taken from the bottom of the U-shaped meniscus. Try to read the initial volume to one place beyond the markings on the cylinder. In the case of a 25-ml graduated cylinder, there is a line that goes completely around the cylinder every 5 ml, a line every 1 ml, and an even smaller line every 0.5 ml. By carefully looking at the bottom of the U-shaped meniscus, you should be able to tell whether the liquid level is half-way between to lines, or a quarter of the way down from a line, or a quarter of the way up from a line. By making this type of "guesstimation" the uncertainty in each volume reading would be on the order of ±0.15 ml, which translates to roughly an error of ±0.25 ml in the volume of displaced liquid, which will be calculated as the difference between the final and initial volume readings. (The marking lines on a 10-ml and 100-ml graduated cylinder are different. Be sure to carefully inspect the graduated cylinder to see how the marking lines are spaced.) Now carefully place the solid object into the graduated cylinder. The solid must be completely submerged. Record the volume of the water + solid object on the laboratory Data Sheet for each unknown or item to be studied. In the case of the smaller-sized objects, you may want to use several so that the volume of liquid displaced is large enough to be measured accurately. Remember that there is an experimental uncertainty associated with each experiment measurement, and in today's experiment you want the volume of the liquid that is displaced by the solid objects to be much larger than the experimental uncertainties associated with the initial and volume readings.

DENSITY DETERMINATION OF WOODEN OBJECTS BY VOLUME FORMULA METHOD

The liquid displacement method requires that the density of the solid object must be larger than the density of the displaced liquid, in other words, the object must sink. Wooden objects typically float in water (densities of various woods include: balsa ≈ 0.11 grams/cm^3; cork ≈ 0.20 grams/cm^3; pine ≈ 0.36 grams/cm^3; redwood ≈ 0.40 grams/cm^3; cedar ≈ 0.42 grams/cm^3; walnut ≈ 0.53 grams/cm^3; teak ≈ 0.56 grams/cm^3; maple ≈

0.63 grams/cm^3; oak ≈ 0.65 grams/cm^3; ebony ≈ 1.1 grams/cm^3; and ironwood ≈ 1.30 grams/cm^3. Although we could replace the water with an organic solvent like cyclohexane (density ≈ 0.779 grams/cm^3 at 25 °C) or isopropyl alcohol (density ≈ 0.798 grams/cm^3 at 25 °C), many of the wooden objects would still float. The solid objects that will be studied today have a regular geometric shape, which allows us to calculate the object's volume based on a mathematical formula (see Eqns. 6.2 – 6.5). Using the caliper constructed from a manila folder, and a ruler (1 inch = 2.540 cm) to measure the distance of the caliper gap, determine the dimensions of the rectangular wooden block and the diameter (radius of sphere = 0.5 × diameter of sphere) of the wooden ball. Record the dimensions and masses of both wooden objects on the laboratory Data Sheet. Calculate the volume of the object by mathematical formula. The density is obtained by substituting the mass and calculated volume into Eqn. 6.1.

DENSITY DETERMINATION OF BALL BEARINGS BY LINEAR LEAST SQUARES ANALYSIS

In this part of the experiment you will determine the masses and volumes of different sized ball bearings made from the same material. The volume of the sphere is calculated as: volume = (4/3) π r^3. Use the manila folder caliper and ruler to the measure the diameter of each sphere. See Figure 6.1 for details. Record the diameter and mass of each ball bearing that is to be studied on the laboratory Data Sheet in the section titled "Density Determination of Ball Bearings by Linear Least Squares Analysis". Calculate the volume of each ball bearing. Next, calculate the density of each ball bearing by substituting the experimental mass and calculated volume into Eqn. 6.1. How do the individual values compare?

The density of the ball bearings will now be determined by linear least squares analysis, by first plotting the volume along the x-axis and mass along the y-axis. This graph should be linear, with the numerical value of the slope equal to the density of the material used to make the ball bearings. The slope of the line and corresponding correlation coefficient are to be determined by the method of linear least squares analysis. Unfortunately, this least squares analysis assumes that the largest experimental uncertainties were in the mass, which is probably not the case here. The uncertainty in the mass of the ball bearing is on the order of ± 0.002 or so. For this experiment the uncertainty in the volume was likely larger. Next graph the data with the mass along the x-axis and volume along the y-axis. The graph should again be linear; however, now the slope is the reciprocal of the density.

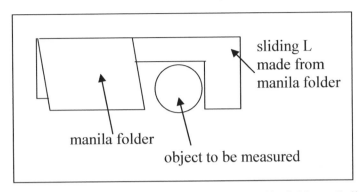

Figure 6.1. Caliper constructed from manila folder. Caliper
is used to measure the diameter of spherical ball bearings

A linear least squares analysis using the Pasco system is performed as follows: Enter the Pasco system by **mouse clicking on the DataStudio icon** that is displayed on the computer screen when the computer is first turned on. You should now see what is called the Welcome to DataStudio screen. **Right mouse click on the Enter Data picture**. On the left-hand side of the screen there should be an x-y data table, and on the right-hand side of the screen there should be an x-y graph. Enter the x-value of the first data point in the first line of the x-column, hit enter, type the y-value of the first data point, enter, type the x-value of the second data point, enter, ..., until all of the data points have been entered. Examine the table carefully for typographical errors. To perform the least squares analysis, **click on the Fit button** at the top of the Graph side toolbar, **scroll down to Linear Fit**, and left click on the mouse. The slope and the intercept of the line, as well as the correlation coefficient, are now displayed inside the box on the computer screen. Record the information in the appropriate space of the laboratory Data Sheet.

To perform the next linear least squares analysis, go up to the file button at the top left-hand corner of the toolbar. **Mouse click on the File Button, scroll down to New Activity, and mouse click**. When the system asks "Should DataStudio save this activity? Click no. You should now be back to the Welcome to DataStudio screen where you began the first least squares analysis calculation. The data table for the first least squares is lost when you exit and re-enter the system.

[If you want to still have access to the first data table, you can bring down an empty data table by pointing the **mouse cursor to the Experiment key on the top toolbar, mouse clicking** and the **scrolling down until you reach New Empty Data Table. Mouse click** and an empty data table appears on the left-hand side of the screen.

Type in the mass (x-column) and volume (y-column) data. The graph will not be automatically generated. To get the graph to appear, you need to **mouse click on the Summary button. Double left-hand mouse click on graph under the Display window**. A new screen titled Choose Data Source will open up. **Move the mouse cursor to the second Editable data line and click**. The graph with the new data should appear. From here go up to the **Fit button and mouse click**. The data entered for the volume (x-axis) and mass (y-axis) is stored in the first Editable data line, and can be retrieved by mouse clicking on it while the Choose Data Source Window is open.]

Microsoft Excel Instructions for Performing Linear Least Squares Regression Analyses: Generally you should have ample time during the three-hour laboratory period to perform all of the required linear least squares analyses using the built-in Pasco software. On a rare occasion the laboratory class may end before all students have finished the least squares calculations, or sometimes students may discover while finishing the laboratory write-up at home, that they forgot to record all of the necessary statistical information on their laboratory Data Sheet. If this should happen to you, Microsoft Excel is capable of doing a linear least squares analysis. The general computer access laboratories across campus should have Microsoft Excel as part of the standard Microsft Windows bundle of software. The Excel spreadsheet will appear as a grid of cells, which are identified by both a column alphabetical letter and row numerical value. Lets assume that you had 5 experimental data points on which you wanted to perform a linear least squares analyses. In cells A5 through A9 enter the numerical x-values of the five data points. In cells B5 through B9 enter the numerical y-values. Once all of the data points are entered, you need to access the built-in statistical package. In cell E1 type- **=INDEX(LINEST(B5:B9,A5:A9),1)** . Hit **ENTER** – by typing this command you have instructed the program to calculate the slope. Note that in the syntax – the y-array of experimental data is entered first, followed by the x-array. The numerical value of the slope should appear in cell E1. To calculate the intercept type in cell E2- **=INDEX(LINEST(B5:B9,A5:A9),2)** . Again hit ENTER. The numerical value of the y-intercept should appear in cell E2. Finally, to get the correlation coefficient, type in cell E3- **=CORREL(B5:B9,A5:A9)** . When you hit ENTER the correlation coefficient should appear in cell E3. The number of data points can be increased. For example, if one had 9 data points, instead of 5, the one-dimensional x- and y-arrays would go to A13 and B13, respectively.

DATA SHEET – EXPERIMENT 6

ball bearings

Name: _____

Density Determination by Liquid Displacement

Unknown Element A:

1. Mass of unknown element A, g: _____

2. Volume of water in graduated cylinder, ml: _____

3. Volume of water + Unknown A in graduated cylinder, ml: _____

4. Volume of element A, ml: _____

5. Density of Unknown A, grams/ml: _____

6. Guessed identity of Unknown A (see Appendix C):   _____

Unknown Element B:

7. Mass of unknown element B, g: _____

8. Volume of water in graduated cylinder, ml: _____

9. Volume of water + Unknown B in graduated cylinder, ml: _____

10. Volume of element B, ml: _____

11. Density of Unknown B, grams/ml: _____

12. Guessed identity of Unknown B (see Appendix C):   _____

Fishing Weights:

13. Mass of fishing weights, g: _____

14. Volume of water in graduated cylinder, ml: _____

15. Volume of water + fishing weights in graduated cylinder, ml: _____

16. Volume of fishing weights, ml: _____

17. Density of fishing weights, grams/ml: _____

18. Based on the measured density, what elements do you think
    were admixed to make the fishing weights?:   _____

Post-1982 Pennies:

19. Mass of post-1982 pennies, g: _____

20. Average mass of a single post-1982 penny, g: _____
    US Treasury Department gives the mass of post-1982
    penny as 2.50 grams

21. Volume of water in graduated cylinder, ml: _____

22. Volume of water + post-1982 pennies in graduated cylinder, ml: _____

23. Volume of post-1982 pennies, ml: _____

24. Density of post-1982 pennies, grams/ml: _____

25. Guessed composition of a post-1982 penny based on density: _____

Pre-1982 Pennies:

26. Mass of pre-1982 pennies, g: _____

27. Average mass of a single pre-1982 penny, g: _____
    US Treasury Department gives the mass of pre-1982
    penny as 3.11 grams

28. Volume of water in graduated cylinder, ml: _____

29. Volume of water + pre-1982 pennies in graduated cylinder, ml: _____

30. Volume of pre-1982 pennies, ml: _____

31. Density of pre-1982 pennies, grams/ml: _____

32. Guessed composition of a pre-1982 penny based on density: _____

Density Determination by Volume Formula Calculation

Wooden rectangular block

33. Volume of wooden block, length × width × height, cm^3: _____

34. Mass of wooden block, grams: _____

35. Density of wooden block, grams/cm^3: _____

36. What kind of wood might the block be made of: _____

Wooden ball

37. Diameter of ball in centimeters, 1 inch = 2.54 cm, cm: _____

38. Volume of wooden ball, $V = (4/3) \pi r^3$, cm^3: _____

39. Density of wooden ball, $grams/cm^3$: _____

40. What kind of wood might the ball be made from: _____

Density Determination of Ball Bearings by Linear Least Squares Analysis

Experimental Data

Diameter inches	Radius cm	Volume $V=(4/3) \pi r^3$, cm^3	Mass grams	Calculated Density $grams/cm^3$
_____	_____	_____	_____	_____
_____	_____	_____	_____	_____
_____	_____	_____	_____	_____
_____	_____	_____	_____	_____
_____	_____	_____	_____	_____
_____	_____	_____	_____	_____
_____	_____	_____	_____	_____
_____	_____	_____	_____	_____
_____	_____	_____	_____	_____

Calculated Results

41. Density of ball bearings, arithmetic average
    of individual ball bearing densities: _____

42. Density of ball bearings from slope of mass (y-axis)
    versus volume (x-axis) linear least squares analysis: _____

43. Correlation coefficient for mass (x-axis) versus volume (y-axis) graph: _____

44. Density of ball bearings calculated as reciprocal of slope of volume
    (y-axis) versus mass (x-axis) linear least squares analysis: _____

45. Correlation coefficient for mass (x-axis) versus volume (y-axis) graph: _____

The density of the steel ball bearings was determined using two linear least squares analysis. In the first analysis the data was entered as (x,y)-ordered pairs of (volume,mass). In the second analysis, the data was ordered as (mass,volume)-ordered pairs. Which of the two data treatments is more appropriate for the experimental data that you measured? Explain you answer in essay style format.

EXPERIMENT 7: GAS LAWS – VERIFICATION OF BOYLE'S LAW, CHARLES' LAW AND AVOGADRO'S LAW

INTRODUCTION

Gas samples have been studied for hundreds of years, and the properties that all gases display have been summarized into laws (or mathematical relationships if you prefer), which are named for their discover. As you will learn in today's laboratory experiment, the properties of volume, pressure, temperature and number of moles for a gas sample are not independent of each other. The properties are inter-related. One sees the relationship between the number of moles of gas and its pressure and volume every time that one fills up a football, basketball or bicycle tire. A bicycle tire inflates (pressure and volume increase) when gas is added, and the tire deflates when gas is removed. One has also likely observed the relationship between temperature and volume (and pressure). A basketball fully inflated indoors does not appear as inflated when it is left outdoors on a cold winter day.

BOYLE'S LAW DETERMINATION

Quantitative measurements on gases date back to as early as the mid to late 1600s when Robert Boyle investigated the relationship between the volume occupied by a trapped gas in a J-shaped glass tube and the total pressure applied against the gas by a mercury column plus the ambient atmospheric pressure. Boyle observed that the volume of trapped gas was

$$\text{volume of gas} = \text{constant(T,moles)} \times (1/\text{pressure applied}) \tag{7.1}$$

inversely proportional to the total pressure applied, irrespective of the nature of the gas. In other words, doubling the total pressure on the trapped gas reduced its volume by a factor of two. Tripling the applied pressure reduced the volume to one-third of its original value, *etc.* Since the experiments were conducted at a constant temperature and fixed amount of trapped gas, the numerical value of the proportionality constant contains any functional dependence that temperature and amount (number of moles) might have on the gas volume

In this part of the laboratory exercise you will be making a similar series of volume-pressure measurements, though using modern instrumentation. Pressure will be measured using the Pasco pressure transducer. The system's built-in software allows one to examine the experimental values for possible linear relationships. Enter the Pasco

77

System by **double left-hand mouse clicking on the DataStudio icon** on the computer screen. When the screen saying "How would you like to use DataStudio" appears, **mouse click on Create Experiment**. Connect the Pasco Absolute Temperature-Pressure Sensor to the PowerLink unit. The system should automatically recognize what accessories are attached. If for some reason the system does not recognize that the Absolute Temperature-Pressure Sensor is attached, disconnect and reconnect the sensor to the PowerLink unit. If this does not work, exit the system and re-enter by mouse clicking on the DataStudio icon. There really is no calibration of the pressure sensor *per se*. You will have to change the units of pressure to atmosphere (the default unit is kPa). To change the displayed unit to atmospheres **mouse click on the Setup button on the top tool bar**. A Setup window should appear indicating that the Absolute Pressure sensor is attached. To the right Absolute Pressure is a shaded box with kPa ▼. **Mouse click on the box**, and then **scroll down to atmospheres**. **Mouse click when the atmospheres line is highlighted**. Close the Setup window and you are ready to start taking measurements.

Trap a sample of air in a 60-ml plastic syringe, and connect the syringe to the Pasco system using a short piece of Tygon tubing. The tubing needs to be short, otherwise the volume inside of the tubing contributes to the volume of the trapped gas. Record the volume of the trapped gas, and the pressure that is displayed on the computer screen on the laboratory Data Sheet. Now push the plastic plunger in to the 50-ml mark and record the new pressure. Continue measuring pressures at the gas volumes listed on the Data Sheet. Once you are finished making the measurements, you need to access the system's built-in software to generate a graph of volume (y-axis) versus pressure (x-axis), a graph of volume (y-axis) versus pressure^2 (x-axis), a graph of volume (y-axis) versus $\sqrt{pressure}$ (x-axis), and finally a graph of volume versus 1/pressure (x-axis). Probably the fastest way to accomplish this task is to use a hand calculator to compute the values of pressure^2, $\sqrt{pressure}$ and 1/pressure for each measured data point. Write the values down on a sheet of paper. If you now enter the data points for each graph in order, starting with the smallest (or largest) x-value (or y-value), the program will automatically generate the respective graphs as the data points are entered. Sketch the graphs on the Data Sheet in the space provided. Which graph was linear? For the linear graph record the equation of the straight line and the correlation coefficient on the Data Sheet.

Does the linear relationship hold for other gases, or is the relationship unique to air? Perform the same set of volume-pressure measurements using "natural" gas. Connect your syringe to the laboratory's "natural" gas supply outlet using the rubber tubing that has been provided for this purpose. Open the gas valve and slowly fill the syringe at least three-fourths full of gas by slowly drawing back the plunger. Turn off the gas valve and disconnect the syringe from the rubber tubing momentarily to empty the syringe by pushing the plunger fully in. This step should purge the syringe of most of the air that may have remained in the syringe. Reconnect the rubber tubing to the syringe, reopen the gas valve, and slowly pull the plunger back until it is a little bit past the 60-ml mark. Turn off the gas valve and disconnect the syringe from the rubber tubing. You are now ready to make the pressure measurements. Gently push the plunger forward to the 60-ml mark, connect the syringe to the Pasco system, and record the gas volumes and pressures as before on the Data Sheet. Once the measurements are finished, access the built-in software to generate a graph of volume (y-axis) versus pressure (x-axis), a graph of volume (y-axis) versus pressure^2 (x-axis), a graph of volume (y-axis) versus $\sqrt{pressure}$ (x-axis), and finally a graph of volume versus 1/pressure (x-axis). Sketch the graphs on the Data Sheet in the space provided. Which graph was linear? For the linear graph record the equation of the straight and the correlation coefficient of the Data Sheet. How does the straight line from the air and methane set of measurements compare?

Ask the TA whether or not you have time to perform a third set of volume-pressure measurements. By now you should be fairly proficient in operating the built-in software. If a third set of measurements is to be made, trap a sample of carbon dioxide by syringing the vapor next to a piece of dry ice. Purge the methane from the syringe by first filling the syringe about two-thirds full with carbon dioxide gas and then pushing the plunger all the way in to empty. Now trap the sample of carbon dioxide. The sample will likely be a little colder than room temperature since it was next to the dry ice. Wait a few minutes for the gas to warm up before connecting the syringe to the Pasco system. Just before connecting the syringe to the Pasco system, push the plunger in to the 60-ml mark. Measure the volumes and pressures as before, and enter the numerical values on your Data Sheet. Once all of the data is entered, access the built-in software to generate a graph of volume (y-axis) versus pressure (x-axis), a graph of volume (y-axis) versus pressure^2 (x-axis), a graph of volume (y-axis) versus $\sqrt{pressure}$ (x-axis), and finally a graph of volume versus 1/pressure (x-axis). Sketch the graphs on the Data Sheet in the space provided. Which graph was linear? For the linear graph record the equation of the straight and the correlation coefficient of the Data Sheet. How does the carbon dioxide line compare to the lines from the air and methane measurements?

CHARLES' LAW DETERMINATION

The mathematical relationship between the volume of a gas and its temperature was not discovered until 1987, when French scientist Jacques Charles found that the volume of a fixed amount of trapped gas:

79

$$\text{volume of gas} = \text{slope} \times \text{temperature of gas} + \text{intercept} \qquad (7.2)$$

increased linearly with temperature. The Kelvin temperature scale was not proposed until the year of 1848 by William Thomson (a British physicist whose title was Lord Kelvin), so the original statement of Charles' law was quite different than the "modern" version that appears in today's introductory level general chemistry textbooks.

As part of the experimental measurements you will verify the linear relationship given by Eqn. 7.2. You will also perform a mathematical technique called "transformation of origin", which will move the graph's origin to a new location so that the observed straight line from the volume-versus-temperature curve passes through the new origin. The transformation-of-origin technique is accomplished by first obtaining the mathematical equation for the straight line

$$\text{volume of gas} = (\text{number filled in for slope}) \times \text{temperature (in }°C) +$$

$$\text{number filled in for intercept} \qquad (7.3)$$

through linear least-squares regression analysis, and then calculating the temperature (in °C) that corresponds to a gas volume of zero. Set volume of gas = 0 in Eqn. 7.3, and solve for temperature (in °C). The temperature that is calculated for volume of gas = 0 tells you how many degrees Celsius the origin needs to move to the left in order for the straight line to pass through the origin. From your general chemistry lecture course you will probably already the answer. Very careful temperature-volume measurements gave a value of –273.15 °C, which is the value in the Kelvin <------> Celsius (Centigrade) conversion. To get back to 0 °C from the new origin of 0 Kelvin, one needs to add 273.15 to the Centigrade (or Celsius) temperature values. It is very doubtful that you will calculate a numerical value of –273.15 °C as there is quite a large extrapolation between the lowest temperature that you will use and absolute zero.

To perform this part of the experiment fill the smaller syringe (10 ml syringe or 15 ml syringe) about 75 % with air. The syringes will have their ends melt sealed. Several different temperature baths will be used. There will be a hot water bath (a beaker of hot water); an ice-water bath; a salt water-ice bath; and an alcohol-dry ice (or acetone-dry ice) bath, in addition to the ambient room temperature. This should give 5 experimental volume-temperature data points for the least squares regression analysis.

Measure the ambient room temperature by placing the thermometer next to the syringe, and record the temperature (room temperature) and volume of trapped gas in the syringe at room temperature on your laboratory Data Sheet in the section titled "Charles' Law Measurement".

WARNING NOTE – DO NOT TRY TO MEASURE THE TEMPERATURE ALCOHOL-DRY ICE (OR ACETONE-DRY ICE) BATH. NEITHER THE

GLASS THERMOMETER NOR PASCO TEMPERAURE PROBES ARE DESIGNED FOR THIS LOW OF TEMPERATURE MEASUREMENT. Instead, use – 65 °C for the temperature of the alcohol-dry ice (or acetone-dry ice) bath. Typical measured values of various solvent-dry ice baths are: - 63 °C for acetone-dry ice; -68 °C for isopropanol-dry ice; and –71 °C for ethanol-dry ice.

Now place the syringe in the hot water bath, and record the bath temperature and gas volume on your Data Sheet after thermal equilibrium is achieved. (This usually takes two to three minutes). Next put the syringe in the ice-water bath, and record the bath temperature and gas volume on your Data Sheet thermal equilibrium is achieved (two to three minutes). Next put the syringe in the salt water-ice bath, and record the bath temperature and gas volume on your Data Sheet after two or three minutes have past. The salt water-ice bath is prepared by adding about a palm-full of salt to the ice-water bath. (Use a liberal amount of salt. A small amount of salt does not lower the bath temperature very much.)

Finally place the syringe in the alcohol-dry ice (or acetone-dry ice) bath, and measure only the volume of gas. **DO NOT TRY TO MEASURE THE TEMPERAURE OF THIS BATH.** Use –65 °C for this temperature. Once all of the volume-temperature measurements are complete, access the built-in software to generate a graph of volume (y-axis) versus temperature (x-axis). Write down equation of the line and the correlation coefficient. Using the slope and intercept from the linear regression, calculate the numerical value of absolute zero (see Eqn. 7.3 for details of this calculation). How does your calculated value compare to the accepted value of –273 °C?

AVOGADRO'S LAW DETERMINATION

This part of Experiment 7 examines how the volume of trapped gas depends on the amount (*e.g.*, number of moles) of gas. The temperature and volume are held constant. The mathematical relationship between the quantity of a gas and its volume follows from the work of Joseph Louis Gay-Lussac and Amadeo Avogadro. In 1808 Gay-Lussac observed that, at a given temperature and pressure, the volume of reacting gases are always in the ratios of small numbers. Today, the reaction of nitrogen gas with hydrogen gas is written in terms of

1 mole of $N_{2(gas)}$ + 3 moles of $H_{2(gas)}$ ------> 2 moles of $NH_{3(gas)}$

the number of moles of reactants and products. In Gay-Lussac's time, the reaction would be expressed

1 liter of $N_{2(gas)}$ + 3 liters of $H_{2(gas)}$ ------> 2 liters of $NH_{3(gas)}$

in terms of volumes. Three years later Avogadro explained Gay-Lussac's observations concerning combining gas volumes by hypothesizing that equal volumes of gases at the same temperature and pressure contain equal number of molecules. Avogadro's law directly follows from Avogadro's hypothesis. Avogadro's law states that the volume of a gas is

$$\text{volume of gas} = \text{constant}(T,P) \times \text{number of moles of gas} \qquad (7.4)$$

directly proportional to the number of moles of gas, with the temperature and pressure held constant. The numerical value of the constant in Eqn. 7.4 does depend on temperature and pressure, as indicated by "(T,P)". A plot of volume (y-axis) versus number of moles (x-axis) should be linear, with the line going through the origin.

The linear relationship will be verified by measuring the amount of gas that is trapped in different volumes. A 60-ml plastic syringe (Luer-lok syringe with latex syringe cap) has been modified for performing this part of the experiment. On the plunger three small holes (nail stops) have been drilled at a 45° angle to the "plastic rib". The holes are drilled to correspond to syringe volumes of 30 ml, 40 ml and 50 ml. The diameter of the holes should be such that a 2-inch finishing nail snuggly fits into the hole.

Determine the mass of the syringe at the first nail stop by placing the latex cap on the syringe, and pulling the plunger back just past the first nail stop. Put the finishing nail into the first hole. The plunger should now be allowed to move slowly forward, until it is stopped by the nail that was just inserted. The plunger may be hard to pull back as a "vacuum" is being formed in the syringe. Have your lab partner help you. Now place the syringe on the balance in an upright position (standing on the handle of the plunger) and weigh. Record the mass of the evacuated syringe at the first nail stop on the laboratory Data Sheet. Repeat this step for the second and third nail stop positions.

It is now time to trap and weigh the gas samples. Large molar mass gases work best for this experiment. Air and carbon dioxide should both be readily available. (Carbon dioxide (dry ice) was used in the Charles law determination for the –65 °C temperature.) If air is used, remove the syringe cap, pull the plunger back just passed the first nail stop, insert the finishing nail, and gently push the plunger move forward until it is stopped by the inserted nail. Put the latex cap on the syringe. Reweigh the syringe and trapped gas. Record the mass and volume of trapped gas on the Data Sheet. For this experiment, the volume of gas is measured using the markings on the syringe. [If greater accuracy is desired, one can fill the syringe with water, and record the mass of water contained in the syringe at each nail stop. The volume is calculated from the density of water, *e.g.*, each gram of water corresponds to 1 ml. Wait until the end of the experiment if you decide to determine the volume in this fashion. It will take awhile for the syringe to dry thoroughly, and you still have two more trapped gas measurements to make.]

You now need to trap and weigh air at both the second and third nail stops. Be sure to record the volume of gas trapped and the mass of the syringe + cap + nail + trapped gas on the Data Sheet in the second titled "Avogadro's Law Determination." The molar mass of air ≈ 28.96 grams/mole. Even though air is not a pure compound, the composition of air is essentially constant at 78.09 % $N_{2(gas)}$, 20.95 % $O_{2(gas)}$, 0.93 % $Ar_{(gas)}$ and 0.03 % $CO_{2(gas)}$.

If you decide to trap carbon dioxide gas, rather than air, place the tip of the syringe close to the evaporating dry ice. Fill the syringe and expel the gas a couple of times to remove any trapped air. Now trap the gas sample by pulling the plunger back past the nail stop, insert the nail stop, and gently push the plunger forward until it is stopped by the nail. The trapped gas sample will likely be colder than room temperature. Leave the latex cap off until the barrel feels like it is at room temperature. Now put the cap on the syringe, and weigh. Record the mass and volume of trapped gas on the Data Sheet. (If you use carbon dioxide by sure to note this on the Data Sheet as well so that the information will be at hand when you calculate the number of moles of trapped gas.) Repeat the trapping and weighing steps for the second and third mole stops.

Plot the volume of the trapped gas (y-axis) versus the number of moles of trapped gas (x-axis) to verify that Avogadro's law was followed. Include the origin as a fourth data point. The volume of the trapped gas is zero when there is no trapped gas. On the Data Sheet record the equation of the straight line and the correlation coefficient. Was a linear relationship obtained?

IDEAL GAS LAW

You have now experimentally verified

Boyle's law:	volume ∝ 1/pressure	T and n held constant
Charles' law:	volume ∝ absolute temperature	P and n held constant
Avogadro's law:	volume ∝ number of moles	P and T held constant

Explain in the space provided on the Data Sheet how the three individual laws are combined to give the Ideal Gas law. (Hint: the explanation was given in the laboratory lecture.) Strictly speaking the ideal gas law equation is a very useful mathematical description for gases; however, all real gases fail to obey the relationship to some degree. Careful experimental measurements have shown that the departure from ideality is different for each gas, and nonideal behavior generally increases with increasing pressure and decreasing temperatures. The deviation results from molecular interactions between neighboring gas molecules that become more prevalent at high pressures, and from the fact that the gas molecules do have finite volumes.

DATA SHEET – EXPERIMENT 7

Robert Boyle

Name: _____

Boyle's Law Measurement

Syringe mark	Air		Natural gas	
	Volume (ml)	Pressure (atm)	Volume (ml)	Pressure (atm)
60-ml mark	_____	_____	_____	_____
50-ml mark	_____	_____	_____	_____
40 ml mark	_____	_____	_____	_____
30 ml mark	_____	_____	_____	_____
25 ml mark	_____	_____	_____	_____
20 ml mark	_____	_____	_____	_____

Syringe mark	Carbon dioxide		Another gas	
	Volume (ml)	Pressure (atm)	Volume (ml)	Pressure (atm)
60-ml mark	_____	_____	_____	_____
50-ml mark	_____	_____	_____	_____
40 ml mark	_____	_____	_____	_____
30 ml mark	_____	_____	_____	_____
25 ml mark	_____	_____	_____	_____
20 ml mark	_____	_____	_____	_____

Boyle's Law Graphs – for Air

Volume versus Pressure   Volume versus pressure^2

Volume versus $\sqrt{pressure}$ Volume versus 1/pressure

Boyle's Law Graphs – for Natural Gas

Volume versus Pressure   Volume versus pressure^2

Volume versus $\sqrt{pressure}$ Volume versus 1/pressure

Charles' Law Measurement

Type of Bath	Temperature (°)	Volume (liters)
Room temperature	_____	_____
Hot water	_____	_____
Water-ice	_____	_____
Salt water-ice	_____	_____
Alcohol-dry ice	-65 °C	_____

1. Sketch the plot of volume of gas versus temperature (°C)

2. Equation for volume versus temperature (°C): _____

3. Correlation coefficient for equation: _____

4. Calculated value for absolute zero, °C: _____

Avogadro's Law Measurement

	1st Nail Stop	2nd Nail Stop	3rd Nail Stop
Mass of the evacuated syringe, g:	_____	_____	_____
Mass of syringe + trapped gas, g:	_____	_____	_____
Mass of trapped gas, g:	_____	_____	_____
Volume of trapped gas, ml:	_____	_____	_____
Moles of trapped gas:	_____	_____	_____

Equation of volume of gas versus moles of gas: _____

Correlation coefficient: _____

Ideal Gas Law Derivation

Explain in a fair amount of detail how the three individual laws are combined to give the Ideal Gas law. (If necessary use additional paper to explain your answer.)

Towards the end of the 1600's the French physicist Guilluame Amontons noted that for any trapped gas whose volume and mass are held constant, the rise in temperature produces the same increase in pressure. The experimental values gave

Pressure = slope × Temperature (in °C) + intercept

a linear relationship between pressure and temperature. Several years later, in 1779 Joseph Lambert defined absolute zero as the temperature at which the pressure of a gas becomes zero when a plot of gas pressure versus temperature is extrapolated to $P_{gas} = 0$. (See L. H. Adcock, *J. Chem. Educ.*, **75**, 1567-1568 (1998) for additional explanation.) How would the value of absolute zero defined in this manner theoretically compare to the value based on Charles' law?

EXPERIMENT 8: DETERMINATION OF COOLING CURVES FOR PURE SUBSTANCES AND MIXTURES

INTRODUCTION

In today's laboratory experiment you will study the solid-to-liquid phase transition for several "pure" substances and for several mixtures using the cooling-curve method. Before the actual experimental procedure is presented, lets discuss the much broader issue of what happens to a substance as heat is added and/or removed. For example, what happens when a block of ice is heated at a slow, constant rate an initial temperature of –50 °C to a final temperature of 110 °C. The block of ice is placed in a sealed container so that the system is a thermodynamically closed system in that no mass change occurs. The heat that is first added goes entirely towards raising the temperature of the ice. The temperature of the ice increases linearly (see line A in Figure 8.1). The slope of the line has units of °C/kJ, which is the reciprocal of the units of heat capacity.

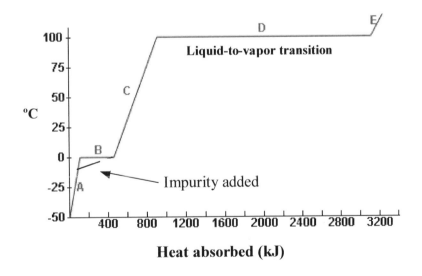

Figure 8.1. Heating curve for observed for the heating of water (initially at –50 °C) to a final temperature of slightly above 100 °C. The horizontal regions in the curve correspond to phase transitions. Region B is for the solid-to-liquid phase transition, and Region D corresponds to the liquid-to-vapor transition.

(Remember the heating rate is constant, *e.g.* heat ∝ time) In other words, the slope of line A is the reciprocal of the heat capacity of the block of ice. At the molecular level the heat that is absorbed makes the hydrogen and oxygen atoms in the water molecule vibrate more extensively about their average positions in the crystalline lattice.

Examination of Figure 8.1 further reveals that the temperature levels off at 0 °C, despite the fact that heat continues to be applied. The heat that is absorbed goes into disrupting hydrogen bonds and other intermolecular forces rather than into increasing the temperature, as indicated by the plateau at 0 °C on the hearting curve (region B). At this temperature – called the melting point temperature – solid and liquid can coexist in equilibrium as molecules break free from their positions in the ice crystals and enter the liquid phase. Once all solid has melted, and not before, does the absorbed heat cause the temperature to rise. The amount of heat required to overcome the intermolecular forces to convert a solid into a liquid is called the enthalpy of fusion (often simply called the heat of fusion). Region C of the heating curve involves a temperature increase, and as you might suspect, the slope of the line is related to the heat capacity of the liquid water sample. Specifically, the slope of line C is the reciprocal of the heat capacity of liquid water sample. The specific heat, which is an intensive property, is obtained by dividing the heat capacity of the liquid sample by its mass. For most substances, the specific heat of the solid and liquid phases are different; ice near 0 °C has a specific heat of SpHt = 2.061 J/(gram-°C), and liquid water near room temperature has a specific heat of SpHt = 4.184 J/(gram-°C). At the molecular level liquid water uses the heat that is absorbed to increase the vibrational motion of the hydrogen and oxygen atoms, and to increase the molecule's rotational motion about its center of mass. At 100 °C a second temperature plateau (region D in Figure 8.1) is observed as the liquid-to-vapor phase transition occurs. The quantity of heat to convert a liquid to vapor is called the enthalpy of vaporization (or heat of vaporization). Examination of Figure 8.1 reveals that the length of region D is much longer than the length of region B, in fact it is about 6.5 to 7 times longer. This means that the enthalpy of vaporization of water is much larger than the enthalpy of fusion of water; about 6.5 to 7 times larger as reflected by the experimental values of $\Delta H_{fusion} = 6.01$ kJ/mole versus $\Delta H_{vaporization} = 40.67$ kJ/mole. Once all liquid water has evaporated, the heat that is absorbed goes toward increasing the temperature (region E).

A cooling curve for water is the process in reverse order. One starts at 110 °C (region E) and then observes the changes in temperature and phase transitions that occur as heat is removed until the final ice temperature of –50 °C is reached. From an experimental standpoint it is easier to generate a cooling curve in the introductory general chemistry laboratory as opposed to trying to measure a heating curve. Heat can be easily removed from a warm object by simply placing the object in a colder temperature bath, such as in an ice-water bath.

Now lets go back and discuss what happens to a salt-water block of ice as the block is heated from an initial temperature of – 50 °C to slightly above the melting point temperature. We do not need to go all the way to the liquid-to-vapor phase transition since today's experiment will involve cooling a warm liquid to just past the solidification temperature. The heat that is first added goes entirely towards raising the temperature of the salt-water ice block. The temperature of the ice block increases linearly as before, but with a slightly different slope. The heat capacity of solid water and solid salt water are not the same. The salt-water ice block though begins to melt at lower temperature than 0 °C. You likely already know this from past experience. Road crews put salt on the roads to melt ice. It takes a colder temperature (lower temperature) for the salt-water to refreeze. The solid-to-liquid phase transition is probably not horizontal though, but likely will have a slightly positive slope. In other words, a block of solid salt-water melts not at a single temperature, but rather over a small temperature interval. The solid starts melting at one temperature, and the melting process is complete at a slightly higher temperature. Once all of the solid salt-water ice block has melted, the absorbed heat goes towards increasing the temperature of the liquid salt-water. For illustrational purposes the impurity effect has been exaggerated in Figure 8.1. Addition of a slight impurity to water would not depress the freezing point temperature this much. For example 58.5 grams of sodium chloride added to 1 kilogram of water would lower the freezing point temperature by only 3.72 °C; however, the line for an effect this small was not distinguishable from the line for pure water using the y-axis scale in Figure 8.1.

How might we use this information in a practical sense? Well the measured melting point temperature of a substance can be used to provide an indication of the purity of the compound. For a pure substance, the melting point temperature is very sharp, that is region B is horizontal. A pure substance melts at a single temperature. For an impure compound, the melting process occurs over a temperature range, rather than at a sharp temperature. The compound might start melting several degrees below the temperature at which all melting is complete. In next week's laboratory experiment you will examine the mathematical relationship between the amount of impurity and the lowering of the freezing point/melting point temperature. Freezing point depression is a colligative property. Other colligative properties are boiling point elevation, vapor pressure lowering and osmotic pressure, the latter of which is related to the migration of solvent and other small molecules through a semi-permeable membrane.

One additional observation regarding cooling curves. Sometimes one observes a phenomenon called supercooling. The temperature of the liquid solution goes below the freezing point temperature as indicated in Figure 8.2. Supercooling represents a metastable thermodynamic state, and its occurrence is not easily predicted. A supercooled liquid is often stable for only so long, as the state becomes increasing less stable with decreasing temperature. Very small disturbances like jostling or stirring can cause the supercooled liquid to suddenly crystallize. Past experience has shown that several of the substances that you will be using in today's laboratory experiment are more likely to supercool than others. A word of caution – be sure that crystallization is complete before you stop the cooling curve measurements. Do not simply think that the substance has solidified simply because the temperature is below the melting point.

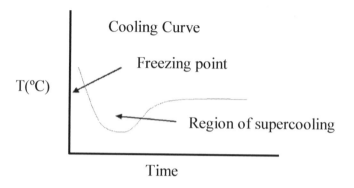

Figure 8.2 Cooling curve of a substance as it passes from the liquid state to the solid solid state. The cooling curve shows a region of supercooling where the temperature momentarily drops below the freezing point temperature. Once nucleation starts the solution temperature increases back to the freezing point temperature.

EXPERIMENTAL PROCEDURE – SETTING UP PASCO SYSTEM AND CALIBRATING THE TEMPERATURE PROBE

Open up the Pasco system by mouse clicking on the Data Studio icon that appears on the computer screen. Now connect the Pasco temperature probe (pH/ORP/ISE Temperature Sensor Pas*Port*) to the PowerLink unit. You will want the **Create Experiment** activity. The system should automatically recognize what probe(s) is (are) connected. After clicking on the Create Experiment activity, **Go to the top toolbar and click on the Setup button.** You will want to **click on the Calibrate button to the right of temperature**. Place the tip of the temperature in an ice-water both (which will serve as the 0 °C reference data), **type in 0 for the Point 1 temperature**, wait a few minutes for thermal equilibrium to be reached, and then **mouse click on the Set button.** The lower reference temperature has now been set. For the higher temperature calibration point, use a heated water bath (above 50 °C). Place the Pasco temperature probe, along with a glass thermometer, into the warm water bath. Measure the temperature of the warm water with the glass thermometer. Now type this value into the box for the Point 2 temperature reading. Wait a few minutes for the temperature probe to achieve thermal equilibrium with the warm water. **Mouse click on the Set button. Now mouse click on OK button.** The Pasco temperature sensor is now calibrated.

PASCO SYSTEM – OPTIONS FOR GRAPHICALLY DISPLAYING TYPES OF EXPERIMENTAL DATA

With the Pasco system you will be measuring the temperature of the solution. If you want to view the data graphically as it is being measured, on the left-hand side of the screen, under the Display options – double left-hand mouse click on graph. Move the mouse pointer to Temperature button and hit OK. You should now have a graph on the right hand side of the screen. The graph can be maximized by placing the mouse pointer over the □ in the upper right-hand corner of the graph box. Once a cooling curve has been generated, you have several options for displaying and manipulating the experimental data.

(a)   If you want to change the x- or y-axis so as to get the entire graph on the screen, **mouse click on the Display button** on the top toolbar, and **scroll down to Settings option, and mouse click.** The axis settings tab lets you scale the axis to a convenient size.

(b)   If you have made multiple temperature runs, and want to select the runs to look at, **mouse click on the Data button on the graph tool bar. Click on what sets of data you want to display graphically.**

(c)   If you want to get numerical values from different spots of the graph, **click on the Display button on the top toolbar**, and **scroll down to Measure option, and mouse click.** You will now have a two-dimensional cursor that you can move across the screen to the various points of the curve. Move the cursor to the spot where you want either the x-value and/or y-value, and both values are displayed off to the side.

EXPERIMENTAL PROCEDURE – COOLING CURVE DETERMINATIONS

The experimental measurements involve determining the cooling curves for the following three "pure" substances and one mixture:

(a)   phenyl salicylate   – freezes in the 40 – 50 °C temperature range;

(b)   benzophenone   – freezes in the 40 – 50 °C temperature range;

93

(c)   acetic acid   - freezes in the 10 – 20 °C temperature range;

(d)   phenyl salicylate + benzophenone mixture 1– prepared by adding about 1.5 grams of phenyl salicylate and 1.5 grams of benzophenone to a test tube. After both chemicals have been added, the test tube is inserted in the hot temperature water bath. The mixture will melt and form a homogeneous solution, which you will then use for the cooling curve measurement.

(e)   phenyl salicylate + benzophenone mixture 2– prepared by adding about 0.75 grams of phenyl salicylate and 2.25 grams of benzophenone to a test tube. After both chemicals have been added, the test tube is inserted in the hot temperature water bath. The mixture will melt and form a homogeneous solution, which you will then use for the cooling curve measurement.

(f)   phenyl salicylate + benzophenone mixture 3– prepared by adding about 2.25 grams of phenyl salicylate and 0.75 grams of benzophenone to a test tube. After both chemicals have been added, the test tube is inserted in the hot temperature water bath. The mixture will melt and form a homogeneous solution, which you will then use for the cooling curve measurement.

The experimental procedure for each of the 6 materials is basically the same. Approximately 3 grams of the material to be analyzed is placed into a test tube. For solid samples you will need to melt the sample once it is in the test tube. This can be easily accomplished by putting the test tube with sample in the hot water bath. You should have already set up the Pasco system to display graphically the experimental data as it is being recorded. If not do so now following the directions in the first paragraph of the preceding section. Once the sample is liquid, place the Pasco temperature into the test tube, with the end of the probe in the liquid. **Click on the Start button on the top tool bar** to start accumulating time versus temperature data. Once the measurement has started promptly place the test tub containing the sample in the ice-water bath. Try to keep the temperature probe from touching the bottom of the test tube. It is best if the probe rests a millimeter or so above the bottom of the test tube. Continue recording experimental data until it is clear from both the graph, and from visual observations that the material in the test tube has indeed solidified. Once the material has frozen continue to record data the solution temperature for a couple of additional minutes so that the entire cooling curve is obtained. To stop recording data, **mouse click on the Stop button** on the top toolbar. You can now examine the experimental data by expanding the x- and y-axis following the directions given in the preceding section. Use the Measure Cursor to record the experimental freezing point temperature. When you get ready to remove the temperature probe from the sample that is frozen, either dissolve the solid in acetone, or warm the test tube in warm water to remelt the solid. **DO NOT ATTEMPT TO REMOVE THE TEMPERATURE PROBE FROM A SAMPLE THAT HAS SOLIDIFIED. YOU MAY BREAK THE PROBE.** Discard the waste compounds and washings as directed by your Teaching Assistant. Organic chemicals should be poured into an organic waste container.

To measure the cooling curve for the next sample, place the temperature probe in the liquid sample, and mouse click on Start button on the top tool bar. The system should generate a new data curve, which would be labeled as "Run 2." Measure the cooling curves for all six materials.

DATA SHEET – EXPERIMENT 8

Name: _____

Cooling Curve for Phenyl Salicylate

1. Melting point temperature, °C: _____

2. Sketch of cooling curve below:

Cooling Curve for Benzophenone

3. Melting point temperature, °C: _____

4. Sketch of cooling curve below:

Cooling Curve for Acetic Acid

5. Melting point temperature, °C: _____

6. Sketch of cooling curve below:

Cooling Curve for Phenyl Salicylate/Benzophenone Mixture 1

7. Melting point temperature, °C: _____

8. Sketch of cooling curve below:

Cooling Curve for Phenyl Salicylate/Benzophenone Mixture 2

9. Melting point temperature, °C: _____

10. Sketch of cooling curve below:

Cooling Curve for Phenyl Salicylate/Benzophenone Mixture 3

11. Melting point temperature, °C: _____

12. Sketch of cooling curve below:

EXPERIMENT 9: DETERMINATION OF MOLAR MASS BY FREEZING POINT DEPRESSION

INTRODUCTION

In Experiment 4 you determined the molar mass of butane, which is the fuel contained in cigarette lighters. The determination required trapping a known mass of butane in a cylinder filled with water. The volume of the trapped gas equaled the volume of water displaced. By substituting the measured values of temperature, pressure, volume and mass of the trapped butane into the ideal gas law, the molar mass of butane was calculated. Such methods are limited in applicability to gases or compounds that readily volatilize near ambient room temperature or at slightly elevated temperatures. Molar masses of nonvolatile compounds, however, cannot be determined in this manner.

Freezing-point depression methods provide a convenient means for determining the molar masses of salts and other nonvolatile compounds. Recall from the cooling curve measurements that when a solute is added to a pure liquid solvent, the freezing point temperature of the resulting solution is lower than that of the pure solvent. The change in the freezing point temperature, ΔT_{fp}, is

$$\Delta T_{fp} = - K_f \times \text{molality solute} \times i \tag{9.1}$$

$$T_{fp,final} - T_{fp,inital} = - K_f \times (\text{moles of solute/mass of solvent in kg}) \times i \tag{9.2}$$

$$\text{moles of solute} = \text{mass of solute/molar mass of solute} \tag{9.3}$$

directly proportional to the molality of dissolved solute. In Eqns. 9.1-9.3, K_f is the molal freezing-point-depression constant (values of K_f for several solvents are listed in Table 9.1) and i is the van't Hoff factor that describes the extent to which electrolytes dissociate in the resulting solution. The numerical value of i equals unity for substances like sugar, which do not ionize in solution. The limiting van't Hoff factor for sodium chloride would be $i = 2$ since NaCl

$$NaCl_{(solid)} \text{ ------> } Na^+_{(aq)} + Cl^-_{(aq)}$$

dissociates into two ions when dissolved in a polar solvent like water. A compound like K_2SO_4, which dissociates into 3 ions in water

$$K_2SO_{4(solid)} \text{ ----> } 2 K^+_{(aq)} + SO_4^{2-}_{(aq)}$$

would have a limiting van't Hoff factor of $i = 3$. Limiting van't Hoff factors are more likely to be realized in dilute solutions. In concentrated electrolyte solutions ion-pair formation occurs and the value of i is reduced below the limiting value. Finally, it should be noted that fractional values of i are possible, and provide valuable information regarding the state of the dissolved solute in solution. For example, a van't Hoff factor of

$i = 1.5$ is halfway between $i = 1$ and $i = 2$. A value of $i = 1.5$ suggests only partial dissociation. Perhaps only 50 % of the solute dissociated into two ions, while the other half remained intact. What would be the dissociation-weighted average value of i under these circumstances? The weighted average would be $i = 1.5$.

Table 9.1 Freezing-point depression and boiling-point elevation constants for several common solvents

Solvent	K_f (kg °C/mole)	K_b (kg °C/mole)	T_{fp} (°C)	T_{bp} (°C)
Acetic acid	3.9	3.07	16.7	118.5
Benzene	5.12	2.53	5.5	80.15
Camphor	37.7	5.95	178.4	208.3
Carbon tetrachloride	29.8	5.02	- 22.3	76.8
Chloroform	4.68	3.63	- 63.5	61.2
Cyclohexane	20.0	2.79	6.5	80.9
Ethanol	1.99	1.22	-114.6	78.4
Phenol	7.27	3.56	42	181.2
Water	1.86	0.512	0.0	100.0

MOLAR MASS DETERMATION OF UNKNOWN

The experimental method used in Experiment 8 for studying cooling curves will be used to determine the freezing point temperatures of cyclohexane (pure solvent), and a solution prepared by dissolving a known mass of an unknown solid in a known mass of cyclohexane. Calibrate the temperature probe as before and use the Pasco system to develop cooling curves that will establish the freezing point temperatures.

Place about 20 ml of cyclohexane (graduated cylinder) into a test tube and determine the freezing point temperature as in Experiment 8. Keep the ice bath full, holding the temperature at about 0 °C. Record the freezing point temperature of cyclohexane on the laboratory Data Sheet in the section titled "Molar Mass of Unknown Determination." Cyclohexane is being used as the solvent because the unknown solid will not dissolve in water. The freezing temperature of pure cyclohexane is listed in Table 9.1, so why is the value being measured here? The answer is quite simple. Any errors in calibrating the Pasco temperature probe will be the same for both the pure solvent and resulting mixture. Calibration errors should cancel (or nearly cancel) when the change in temperature is calculated.

Place a test tube in a beaker and weigh on a top-loading electronic balance. Record the mass on the Data Sheet. Next add 20 ml of cyclohexane (graduated cylinder or pipette) to the test tube and reweigh. Record the mass on the Data Sheet. The mass of cyclohexane is calculated by difference. Next place a piece of weighing paper on the balance pan, and rezero (tare) the balance with the paper on the pan. Place about 0.2 grams of unknown nonelectrolyte solid on the weighing paper. Record the actual mass in the appropriate space on the Data Sheet. Add all of the solid weighed out to the test tube containing cyclohexane. Make sure the solute is completely dissolved before continuing - check for undissolved crystals. Determine the freezing point temperature of the resulting mixture, $T_{fp,soln}$, using the cooling curve procedure of Experiment 8, and record the value on the Data Sheet.

Calculate the molar mass of the unknown solid by substituting the measured values into Eqns. 9.2 and 9.3. Remember that the molal freezing-point-depression constant for cyclohexane is $K_f = 20.0$ (kg °C/mole). In the molar mass computation, $i = 1$ (the unknown compound does not dissociate). The mass of the solvent must be in kilograms.

MOLAR MASS DETERMINATION OF ETHYLENE GLYCOL

Ethylene glycol is soluble in water. Place 20 ml of water (graduated cylinder) into a test tube and determine the freezing point as above. Record the value on the Data Sheet. You will need to use a colder temperature bath, however, as water and the various aqueous solutions that will be studied in this section and in the next two sections will be freezing at 0 °C and below. Prepare an alcohol-dry ice bath as in Experiment 8.

CAUTION: DRY ICE CAN CAUSE SEVERE FROST BITE. HANDLE THE DRY ICE CARFULLY, AND KEEP CONTACT BETWEEN YOURSELF AND THE DRY ICE TO A MINIMUM – ONLY A SECOND OR TWO. Cloth gloves are excellent protection for handling dry ice, or alternatively, paper towels can be used to wrap around pieces of dry ice when handling and when inserting pieces of dry ice into the alcohol bath.

As in the molar mass determination of the unknown solid, accurate masses will be needed for both water (solvent) and ethylene glycol (solute). Ethylene glycol is used as the radiator coolant (antifreeze) in automotive vehicles. Use a test tube and beaker as before. Weigh the empty test tube and beaker, and record the mass on the Data Sheet. Next, add 20 ml of deionized water (graduated cylinder or pipette) and weigh accurately. Record the mass on the Data Sheet. Next add 1 ml of antifreeze (graduated cylinder or pipette) and reweigh to determine the mass of antifreeze by difference. Be sure to record the values on the Data Sheet. Make certain that the ethylene glycol has completely dissolved. Determine the freezing point of the aqueous-ethylene glycol mixture using the Pasco temperature probe and the cooling curve procedure of Experiment 8. Record the freezing point temperature of the solution, $T_{fp,final}$, on the Data Sheet.

Calculate the molar mass of ethylene glycol by substituting the measured quantities into Eqns. 9.2 and 9.3. The molal freezing-point depression constant of water is $K_f = 1.86$ (kg °C/mole). Note: ethylene glycol does not dissociate in water, so the van't Hoff constant is $i = 1$.

Ethylene glycol has one of the following molecular formulas: $C_2H_6O_2$, $C_3H_8O_2$, $C_3H_8O_3$, $C_4H_{10}O_2$, $C_5H_{12}O_2$. Based on the experimental molar mass that you calculated, what is the molecular formula of ethylene glycol?

MOLAR MASS DETERMINATION OF SUGAR

Repeat the ethylene glycol procedure with sugar instead of antifreeze. The pure water freezing point determination does not need to be performed again. Use the value from the ethylene glycol set of measurements. Use 20 ml of water as before and 10 grams of sugar. Sugar is fairly soluble in water, even in cold water. (One should be able to use up to 20 grams of sugar without experiencing problems with solubility.) Be sure to record all of the masses on the Data Sheet. Make certain that all of the sugar has dissolved before continuing. Determine the freezing point temperature of the aqueous-sugar solution, using the cooling curve method. Record $T_{fp,final}$ in the space provided on the Data Sheet.

Calculate the molar mass of sugar by substituting the measured quantities into Eqns. 9.2 and 9.3. Use $i = 1$ for this computation. Sugar does not dissociate in water.

MOLAR MASS DETERMINATION OF SODIUM CHLORIDE

Repeat the ethylene glycol procedure using sodium chloride. The pure water freezing point determination does not need to be performed again. Use the value from the ethylene glycol set of measurements. Use 20 ml of water as before and 4 grams of sodium chloride. Do not use too much sodium chloride – it has limited solubility in cold water. Be sure to record the actual masses of water and sodium chloride used on the Data Sheet. Before making the cooling curve measurement – make certain that all of the sodium chloride has dissolved. Measure the freezing point temperature of the sodium chloride solution, and enter the value on the Data Sheet.

Calculate the molar mass of sodium chloride by substituting the measured values into Eqns. 9.2 and 9.3. Even though you know that sodium chloride readily ionizes in water, still use $i = 1$ for this calculation. How does the calculated value compare to the *true* molar mass of NaCl (molar mass of NaCl = 58.45 grams/mole)?

If one knows the molar mass of the solute through some other experimental method, then freezing-point depression measurements can be used to calculate the van't Hoff factor. Calculate i for NaCl by substituting the measured experimental and known molar mass of NaCl into Eqns. 9.2 and 9.3. What does the calculated value suggest about the dissociation of NaCl?

DATA SHEET – EXPERIMENT 9

antifreeze

Name: _____

Molar Mass of Unknown Determination

1. Freezing point of pure cyclohexane, °C: _____

2. Mass of beaker and test tube, g: _____

3. Mass of beaker and test tube and cyclohexane, g: _____

4. Mass of cyclohexane, g: _____

5. Mass of unknown solid, g: _____

6. Freezing point of cyclohexane + unknown soln, °C: _____

7. Molar mass of unknown solute, g/mole: _____

Molar Mass of Ethylene Glycol Determination

8. Freezing point of deionized water, °C: _____

9. Mass of beaker and test tube, g: _____

10. Mass of beaker and test tube and water, g: _____

11. Mass of water, g: _____

12. Mass of beaker, test tube, water + ethylene glycol, g: _____

13. Mass of ethylene glycol, g: _____

14. Freezing point of water – ethylene glycol soln, °C: _____

15. Molar mass of ethylene glycol, g/mole: _____

16. Molecular formula of ethylene glycol is: _____
    (Based on choices given in laboratory manual)

Molar Mass of Sugar Determination

17. Freezing point of deionized water, °C: _____

18. Mass of beaker and test tube, g: _____

19. Mass of beaker and test tube and water, g: _____

20. Mass of water, g: _____

21. Mass of beaker, test tube, water + sugar, g: _____

22. Mass of sugar, g: _____

23. Freezing point of water – sugar soln, °C: _____

24. Molar mass of sugar, g/mole: _____

Molar Mass of Sodium Chloride Determination

25. Freezing point of deionized water, °C: _____

26. Mass of beaker and test tube, g: _____

27. Mass of beaker and test tube and water, g: _____

28. Mass of water, g: _____

29. Mass of beaker, test tube, water + sodium chloride, g: _____

30. Mass of sodium chloride, g: _____

31. Freezing point of water – sodium chloride soln, °C: _____

32. Molar mass of sodium chloride, g/mole: _____
    (For this calculation assume $i = 1$)

33. Measured van't Hoff factor for sodium chloride: _____

EXPERIMENT 10: INTRODUCTION TO CALORIMETRY – DETERMINATION OF SPECIFIC HEATS OF SOLIDS AND LIQUIDS, AND ENTHALPY OF FUSION OF WATER

INTRODUCTION

A thermochemical equation tells us how much energy is transferred as a chemical reaction or process occurs. By convention, a chemical reaction (or process) is exothermic if heat is released, and the numerical value that describes the heat that is released has a negative sign whenever the value is removed from the balanced chemical reaction/process, *i.e.*, ΔH = -365.6 kJ/mole. Chemical reactions and processes that absorb heat are endothermic, and the accompanying change in the enthalpy has a positive sign when it is not included as part of the balanced chemical reaction/process. For example, the melting of one mole of ice to form one mole of liquid water would be written as

$$H_2O_{(solid)} + 6.01 \text{ kJ} \quad ----> \quad H_2O_{(liq)}$$
$$\Delta H_{fusion} = 6.01 \text{ kJ}$$

The positive value of 6.01 kJ indicates that one must add heat in order to melt ice. Had the process been reversed so as to convert one mole of liquid water into ice

$$H_2O_{(liq)} + \quad ----> \quad H_2O_{(solid)} + 6.01 \text{ kJ}$$
$$\Delta H_{fusion} = -6.01 \text{ kJ}$$

the negative sign in front of the numerical value would indicate that heat is released. Thermochemical equations provide valuable information regarding the heat (enthalpy) "content" of chemical reactants and products.

For many chemical reactions and processes, it is possible to measure experimentally the heat using a calorimeter. A calorimeter is a device that measures heat transfers. Calorimetric measurements can be performed at constant volume or constant pressure. In Experiment 10 a constant pressure calorimeter will be used to measure the specific heat of a metal object, the specific heat of a liquid mixture, and the heat (enthalpy) of melting of ice. Enthalpy changes for chemical reactions will be measured in Experiment 11 once we have learned the basic calorimetric method.

Heat capacity and specific heat are the quantities used to quantitatively describe the ability of a material to absorb heat. Heat capacity is an extensive property, and it is defined as the quantity of heat required in order to increase the temperature of that object by one degree Celsius. Being an extensive property, heat capacity naturally depends on the mass of the object. Specific heat, on the other hand, is an intensive property that is independent of the amount of material. The specific heat is the quantity of heat needed to increase the temperature of gram of a substance by one degree Celsius. For water at 15 °C, the specific heat is 4.184 J/(gram °C). The difference between heat capacity and specific heat is illustrated with the following example. Which would require more heat to raise the temperature of water from 15 °C to 16 °C; a bathtub full of water or a cup of water? A bathtub full of water would require more heat as it contains considerably more water. Here we are dealing with the heat capacities of the two objects, a bathtub of water versus a cup of water. The mass is an important consideration. Now take one gram of water from the bathtub and one gram of water from the cup. Which would require more heat? The amount of heat would be the same as the comparison has now been put on "a per gram basis." Specific heat is the ratio of two extensive properties, *e.g.*, specific heat = heat capacity/mass.

CALIBRATE PASCO TEMPERATURE PROBE

Open up the Pasco system by mouse clicking on the Data Studio icon that appears on the computer screen. Now connect the Pasco temperature probe (pH/ORP/ISE Temperature Sensor Pas*Port*) to the PowerLink unit. You will want the **Create Experiment** activity. The system should automatically recognize what probe(s) is (are) connected. After clicking on the Create Experiment activity, **Go to the top toolbar and click on the Setup button.** You will want to **click on the Calibrate button to the right of temperature**. Place the tip of the temperature in an ice-water both (which will serve as the 0 °C reference data), **type in 0 for the Point 1 temperature**, wait a few minutes for thermal equilibrium to be reached, and then **mouse click on the Set button.** The lower reference temperature has now been set. For the higher temperature calibration point, use a heated water bath (above 80 °C). You will need to use a wide temperature range for this experiment as the metal object for the heat capacity measurement will be placed in nearly boiling water. Place the Pasco temperature probe, along with a glass thermometer, into the hot water bath. Measure the temperature of the warm water with the glass thermometer. Now type this value into the box for the Point 2 temperature reading. Wait a few minutes for the temperature probe to achieve thermal equilibrium with the warm water. **Mouse click on the Set button. Now mouse click on OK button.** The Pasco temperature sensor is now calibrated.

VERIFICATION OF HEAT BALANCE EQUATION

In this part of the experiment we will mix samples of hot and cold water in order to verify that the heat lost during a process equals the heat gained during the process. As a result of mixing heat will be transferred from the hot water to the cold water. Once thermal equilibrium is reached, both of the water samples will be at the same temperature, *e.g.*, $T_{final,cw} = T_{final,hw}$ in Eqns 10.1-10.3 below. The final temperature of the solution can be experimentally measured, and compared to the theoretical value. The following heat balance:

$$\text{heat gained by cold water} = \text{- heat lost by hot water} \tag{10.1}$$

$$(SpHt_{water}) \times \text{mass of cold water} \times (T_{final,cw} - T_{initial,cw}) =$$
$$- (SpHt_{water}) \times \text{mass of hot water} \times (T_{final,\,hw} - T_{initial,\,hw}) \tag{10.2}$$

$$4.184 \text{ J/(mole gram)} \times \text{mass of cold water} \times (T_{final,cw} - T_{initial,cw}) =$$
$$- 4.184 \text{ J/(mole gram)} \times \text{mass of hot water} \times (T_{final,hw} - T_{initial,hw}) \tag{10.3}$$

allows one to calculate what the final temperature of the solution should be after mixing. The negative sign in Eqns. 10.1 – 10.3 is there because heat gained is an endothermic process, whereas heat lost is an exothermic process. Every quantity will be experimentally measured and substituted into Eqn. 10.3, except for the final temperature, $T_{final,cw} = T_{final,hw}$. With a hot plate (or Bunsen burner) you will need to heat 250 ml of deionized water up to a temperature of about 40 °C. While you are doing this, weigh an empty Styrofoam cup on the balance and record the mass of the empty cup on the Data Sheet in the space provided in the section titled "Heat Balance Verification Measurements". Now add about 30 ml of room temperature deionized water to the Styrofoam cup using a graduated cylinder, and reweigh the cup with deionized water added. Record the mass on the Data Sheet. Once the water has heated to about 40 °C, you will need to transfer about 30 ml of the hot water to a graduated cylinder. Before adding the hot water to the Styrofoam cup containing the cold water, you will need to measure the temperature of both the hot- and cold-water solutions using the Pasco temperature probe. (Measure the cold water temperature first and hot water temperature last.) Once you have measured and recorded both temperatures, pour the hot water in the graduated cylinder into the Styrofoam cup containing the cold water. Monitoring the temperature of the solution until thermal equilibrium is achieved. Once the temperature has stabilized to a constant value, record the final temperature on the Data sheet. Should a downward trend be observed in the plateau, take the maximum temperature measured as T_{final}. (Note - If one records the temperature of the solution continuously, the resulting graph of temperature versus time would resemble the thermogram depicted in Figure 10.1. The final temperature and change in temperature, $\Delta T = T_{final} - T_{initial}$, can be obtained from the thermogram by extrapolating the linear portions "prior to mixing" and "after mixing" back to the time of mixing.) After the final temperature is recorded you will need to reweigh the Styrofoam cup so that the mass of the hot water can be determined. Record the mass on the Data sheet in the box labeled "Mass of Styrofoam cup + cold water + hot water". Filling in the initial temperatures of the hot and cold water, and the mass of the hot and cold water into Eqn. 10.3 above, calculate what the final temperature of the solution should have been. How does the calculated value compare with your experimentally determined final temperature?

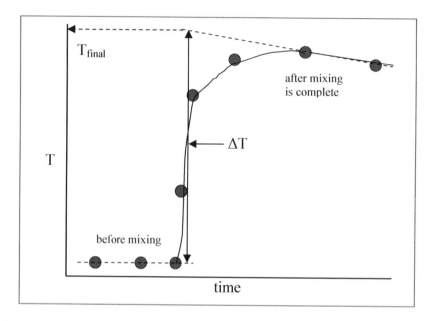

Figure 10.1. Thermogram depicting the variation of solution temperature with respect to time for mixing of cold and hot water. The thermogram that is shown is for adding hot water to cold water. If one were to add cold water to hot water, the solution temperature would have decreased. The solid circles trace the experimental values.

DETERMINATION OF SPECIFIC HEAT OF A METAL

The specific heat of a metal object can be determining the temperature change that occurs when a hot metal object is placed in a solution of water. Here, the heat balance equation is

- heat lost by metal object = heat gained by colder water (10.4)

$- (SpHt_{metal}) \times$ mass of metal object $\times (T_{final,metal} - T_{initial,metal}) =$
$(SpHt_{water}) \times$ mass of hot water $\times (T_{final,water} - T_{initial,water})$ (10.5)

$- (SpHt_{metal}) \times$ mass of metal object $\times (T_{final,metal} - T_{initial,metal}) =$
4.184 J/(mole gram) $\times$ mass of water $\times (T_{final,water} - T_{initial,water})$ (10.6)

As before, when thermal equilibrium is reached, the metal object and the water must be at the same final temperature, $e.g.$, $T_{final,metal} = T_{final,water}$. Every quantity in Eqn. 10.6 will be measured, except for $SpHt_{metal}$ of the metal, which we will calculate. You will need to carefully weigh the specimen of metal to be studied on the electronic balance. Record the

112

mass on the Data Sheet in the section labeled "Specific Heat of Metal Measurements". Weigh an empty Styrofoam cup and record the mass on the Data sheet. Add 100.0 mL of deionized water (with graduated cylinder) to the Styrofoam cup, reweigh the cup with water added, and record the mass in the appropriate space on the Data sheet. Meanwhile, add the metal object to a beaker of hot water (prepared by you using a hot plate or Bunsen burner). Record the temperature of this hot water with a thermometer. Any temperature will do, but a higher temperature will allow better accuracy – water close to the boiling point temperature would be excellent. After the metal has thermally equilibrated in the hot water bath (a couple of minutes), start recording the temperature of the water in the Styrofoam cup using the Pasco Temperature Probe. Remove the metal object from the hot water bath with a pair of tongs, shake quickly to remove excess water, and carefully place the metal object into the calorimeter cup (do all of this promptly so that there is minimum change in the temperature of the metal during the transfer). Just prior to adding the metal object, note the temperature of the water that is being recorded by the Pasco Temperature Probe. This will be the initial temperature of the water solution in Eqn. 10.6. Record $T_{initial,water}$ on your Data sheet. After the temperature of the solution has stabilized, take the final temperature reading and record the numerical value on your Data Sheet. The temperature of the water should rise. Should a downward trend be observed in the plateau, take the maximum temperature measured as $T_{final,water}$. Filling in the initial temperatures of the hot metal object and deionized water, final temperature, and the mass of the metal object and cold water into Eqn. 10.6 above, calculate the specific heat of the metal object.

DETERMINATION OF SPECIFIC HEAT OF A LIQUID MIXTURE

The specific heat of a liquid solution can be experimentally determined by measuring the change in temperature that results when a hot metal object of known specific heat is placed in liquid solution. (The specific heat of a metal object can be obtained by determining the temperature change that occurs when the metal object is placed in a hot solution of water.) Here, the heat balance equation is:

$$- \text{heat lost by metal object} = \text{heat gained by liquid mixture} \qquad (10.7)$$

$$- (SpHt_{metal}) \times \text{mass of metal object} \times (T_{final,metal} - T_{initial,metal}) =$$
$$(SpHt_{liq\,mix}) \times \text{mass of liquid mixture} \times (T_{final,\,liq} - T_{initial,\,liq}) \qquad (10.8)$$
$$- (SpHt_{metal}) \times \text{mass of metal object} \times (T_{final,metal} - T_{initial,metal}) =$$
$$(SpHt_{liq\,mix}) \times \text{mass of liquid mixture} \times (T_{final,liq} - T_{initial,liq}) \qquad (10.9)$$

When thermal equilibrium is reached, the metal object and the liquid solution must be at the same final temperature, *e.g.*, $T_{final,metal} = T_{final,liq}$. Every quantity in Eqn. 10.9 will be measured, except for $SpHt_{liq\ mix}$ of the liquid mixture, which you will calculate. Use the same metal object so that you will know the $SpHt_{metal}$. (The specific heat of the metal object was determined in the preceding section of Experiment 10.) The experimental methodology is basically the same as that used in the determination of the specific heat of a metal, except that water will be replaced by the liquid mixture whose specific heat you wish to determine. (For the laboratory experiment we are using an aqueous-ethylene glycol mixture, the liquid is used as the radiant coolant in automobiles.) You will need to carefully weigh the metal object on the electronic balance. Record the mass on the Data Sheet in the section labeled "Specific Heat of Liquid Mixture Measurements".

Weigh an empty Styrofoam cup and record the mass on the Data sheet. Add 100.0 ml of aqueous-ethylene glycol mixture (with graduated cylinder) to the Styrofoam cup, reweigh the cup with liquid added, and record the mass in the appropriate space on the Data sheet. Meanwhile, add the metal object to a beaker of hot water (prepared by you using a hot plate or Bunsen burner). Record the temperature of this hot water with a thermometer. Any temperature will do, but a higher temperature will allow better accuracy – water close to the boiling point temperature would be excellent. After the metal has thermal equilibrated in the hot water bath (a couple of minutes), start recording the temperature of the aqueous-ethylene glycol mixture in the Styrofoam cup using the Pasco Temperature Probe. Remove the metal object from the hot water bath with a pair of tongs, shake quickly to remove excess water, and carefully place the metal object into the calorimeter cup (do all of this promptly so that there is minimum temperature of the metal during the transfer). Just prior to adding the metal object, note the temperature of the aqueous-ethylene glycol mixture that is being recorded by the Pasco Temperature Probe. This will be the initial temperature of the liquid solution in Eqn. 10.9. Record $T_{initial,liq}$ on your Data sheet. After the temperature of the solution has stabilized, take the final temperature reading and record the numerical value on your Data Sheet. Filling in the initial temperatures of the hot metal object and aqueous-ethylene glycol, final temperature, the mass of the metal object and aqueous-ethylene glycol mixture, and the specific heat of the metal into Eqn. 10.9 above, calculate the specific heat of the aqueous-ethylene glycol that you used.

DETERMINATION OF THE ENTHALPY OF MELTING OF WATER

In this part of the experiment the enthalpy of melting of ice will be experimentally determined by monitoring the temperature change that occurs as ice is added to liquid water. The basic principles are the same; however, the computation is a little bit more involved because ice uses the heat that it absorbs in one of two ways, depending on the temperature that the ice is at. First, if the temperature of the ice is below water's normal melting point temperature, the heat that is absorbed goes towards increasing the temperature of the ice. The temperature of the ice continues to increase until the melting point temperature is reached. At this point in time, the heat that is absorbed by the ice is used to melt the ice. Once all of the ice is melted, and if there is still additional heat that needs to be transferred to reach thermal equilibrium, the additional heat is absorbed by the water that came from the melted ice. The process sounds quite complicated; however, the experimental methodology and computations are fairly simple, provided that one judiciously designs the experiment so that all of the ice melts.

The heat balance that describes the heat transferred when ice is added to warm liquid water is:

$$\text{heat gained by ice} = - \text{heat lost by warm water} \tag{10.10}$$

$$\text{heat used to raise temperature of ice from } T_{initial,ice} \text{ to } 0 \text{ °C} + \text{heat used to melt the ice} +$$
$$\text{heat used to raise temperature of water from melted ice from } 0 \text{ °C to } T_{final} =$$
$$- \text{heat lost by warm water} \tag{10.11}$$

The first and third terms on the left-hand side of Eqn. 10.11 can be calculated using the specific heat of ice and water, respectively. The specific heat of ice is 2.061 J/(gram °C) and the specific heat of liquid water equals 4.184 J/(gram °C). For most substances the heat capacity of the liquid and solid form are different. The second term on the left-hand term involves the enthalpy of melting, which is what we want to calculate. By substituting all of the known and measurable quantities into Eqn. 10.11, we arrive at the following expression:

$$2.061 \times \text{mass of ice} \times (0 - T_{initial,ice}) + \text{mass of ice melted} \times \text{enthalpy of melting of ice}$$
$$(\text{J/gram}) + 4.184 \times \text{mass of water from melted ice} \times (T_{final} - 0) =$$
$$- 4.184 \times \text{mass of warm water} \times (T_{final} - T_{initial,warm\ water}) \tag{10.12}$$

The quantity that we want to calculate is the enthalpy of melting of ice (in units of J/gram), which can be converted to the molar enthalpy of melting by multiplying by the molar mass of H_2O. If all of the ice melts, as should be the case if the directions are followed, then the mass of ice = mass of ice melted = mass of water from melted ice. Otherwise, one would have to determine accurately how much ice actually melted by "fishing out" and "quickly drying" the unmelted ice so that one can calculate the three different "ice" masses needed in Eqn. 10.12.

The experimental procedure for measuring the heat of melting is as follows: Weigh an empty Styrofoam cup and record the mass on the Data Sheet in the section labeled "Enthalpy of Melting of Ice Measurement." Next add 100 ml of deionized water (with graduated cylinder) to the Styrofoam cup and reweigh the cup and its contents. Enter the mass on the Data Sheet in the appropriate spot. Measure and record the initial temperature of the ice using the Pasco temperature probe. Place the temperature probe in the liquid water and record the initial temperature of the water. Now take about 5 grams of ice (wait until the experiment is over to measure the quantity of ice used by difference), blot quickly with a paper towel and add the ice to the Styrofoam cup containing the deionized water. Once thermal equilibrium is reached, the temperature should stabilize. Measure and record the final solution temperature on the Data Sheet. All of the ice should be melted. Now you need to reweigh the Styrofoam cup and contents so that the mass of the melted ice can be calculated by difference. Insert the measured numerical values into Eqn. 10.12. The only quantity that should be unknown at this time is the enthalpy of melting ice, which is what you want to calculate. Calculate the enthalpy of melting of ice, and record the answer in the appropriate place on the laboratory Data Sheet. Calculate the molar enthalpy of melting of ice by multiplying by the molar mass of H_2O. Record the molar enthalpy of melting of ice, in units of kJ/mole, on your Data Sheet. The literature value for the molar enthalpy of fusion of ice is 6.01 kJ/mole. How does your measured value compare to the literature value. Compute the % error, which is defined as

$$\text{Percent error} = 100 \times (\text{exptl. value} - \text{literature value})/\text{literature value} \qquad (10.13)$$

DATA SHEET – EXPERIMENT 10

Name: _____

Heat Balance Verification Measurements:

1. Mass of empty Styrofoam cup, g: _____

2. Mass of Styrofoam cup + cold water, g: _____

3. Mass of cold water, g: _____

4. Mass of Styrofoam cup + cold water + hot water, g: _____

5. Mass of hot water, g: _____

6. Initial Temperature of cold water, °C: _____

7. Initial Temperature of hot water, °C: _____

8. Final Temperature (Experimental), °C: _____

9. Final Temperature (Calculated, Eqn. 10.3), °C: _____

Specific Heat of Metal Measurements:

10. Mass of metal object, g: _____

11. Mass of empty Styrofoam cup, g: _____

12. Mass of Styrofoam cup + deionized water, g: _____

13. Mass of deionized water, g: _____

14. Initial temperature of hot metal object, °C: _____

15. Initial temperature of deionized water, °C: _____

16. Final temperature of deionized water, °C: _____

17. Final temperature of metal object, °C: _____

18. Specific heat of metal, Eqn. 10.6, J/(gram °C): _____

Specific Heat of Liquid Mixture Measurements:

19. Mass of metal object, g: _____

20. Mass of empty Styrofoam cup, g: _____

21. Mass of Styrofoam cup + aqueous-ethylene glycol, g: _____

22. Mass of aqueous-ethylene glycol mixture, g: _____

23. Initial temperature of hot metal object, °C: _____

24. Initial temperature of aqueous-ethylene glycol, °C: _____

25. Final temperature of aqueous-ethylene glycol, °C: _____

26. Final temperature of metal object, °C: _____

27. Specific heat of aqueous-ethylene glycol, J/(gram °C): _____
      (See Eqn. 10.9 for calculation)

Enthalpy of Melting of Ice Measurement:

28. Mass of empty Styrofoam cup, g: _____

29. Mass of Styrofoam cup + deionized water, g: _____

30. Mass of deionized water, g: _____

31. Initial temperature of ice, °C: _____

32. Initial temperature of deionized water, °C: _____

33. Final temperature of solution, °C: _____

34. Mass of Styrofoam cup + deionized water + melted ice, g: _____

35. Mass of ice, g: _____

36. Enthalpy of melting of ice, J/gram: _____

37. Molar enthalpy of melting of ice, kJ/mole: _____

38. Molar enthalpy of melting of ice, lit. value, kJ/mole: _____

39. Percent error: _____

EXPERIMENT 11A: CALORIMETRY II – DETERMATION OF HEAT OF CHEMICAL REACTIONS AND HEAT OF DISSOLUTION

INTRODUCTION

Calorimetric methods are used to determine the enthalpy change that accompanies a chemical reaction. In today's laboratory experiment enthalpies of reaction will be measured for acid-base reaction(s), and for dissolving various electrolyte solutes in water. Each reaction involves bond breakage and formation as the reactants are transformed into products. Bond breaking is an endothermic process, irrespective of whether the bond broken is covalent, or ionic, or hydrogen bond in nature. Bond formation, on the other hand, is exothermic. Heat is released when chemical bonds are formed.

The six specific reactions that are to be studied all take place in aqueous solution. The Styrofoam cup calorimeter will again be used. The calorimeter is not sealed, so the reactions occur under the essentially constant pressure of the atmosphere. Our experience has shown that Styrofoam cups are sufficiently insulated, and that transfer of heat between the solution and its outside surroundings is not a major concern. Styrofoam has a very low thermal conductivity and specific heat. For a good adiabatic calorimeter, the heat gained (or lost) by the solution must be:

$$\text{heat of reaction} = -\text{ heat lost or gained by solution} \qquad (11A.1)$$

equal to, but opposite in sign, as the heat evolved (or gained) by the reaction. The right-hand side of Eqn. 11A.1 is calculated from

$$\text{heat of reaction} = -\text{ (SpHt of solution)} \times \text{(mass of solution)} \times (T_{\text{final,soln}} - T_{\text{initial,soln}})$$
$$(11A.2)$$

the mass of the solution, its specific heat, and the temperature change. For dilute aqueous solutions, the specific heat of the solution will be approximately the same s that of water, 4.184 J/(gram °C).

ENTHALPY OF ACID-BASE REACTION – HCl + NaOH

The enthalpy associated with the reaction of aqueous hydrochloric acid (HCl) and aqueous sodium hydroxide (NaOH) to form water

$$HCl_{(aq)} + NaOH_{(aq)} \longrightarrow H_2O_{(liq)} + NaCl_{(aq)}$$

depends on the amount of chemicals reacted. Enthalpy is an extensive property. Once the enthalpy of reaction is determined for the given amount of reactants consumed, the value is generally scaled up to a per mole basis. In the case of an acid-base reaction, the reported value generally corresponds to the formation of one mole of liquid water.

For this measurement it is imperative that the same amount of hydrochloric acid and sodium solution be used. A Pasco temperature probe is used to measure the initial and final solution temperature. Set the Pasco system up to measure temperature and calibrate the temperature probe. For this experiment you should be able to use an ice-water bath for the lower reference temperature (0 °C) and warm water (in the 40 – 50 °C temperature range) for the upper reference temperature. The instructions for calibrating the temperature probe, and for performing time versus temperature measurements, are found in Experiment 8.

Once this is done weigh an empty Styrofoam cup on a top loading electronic balance. Record the mass of the empty Styrofoam cup on the Data Sheet in the section labeled "Enthalpy Determination for HCl + NaOH reaction". Next add 50.0 ml of 1.0 Molar hydrochloric acid (preferably by pipette, however, a graduated cylinder will do) to the Styrofoam cup and reweigh. Record the mass of the Styrofoam cup + HCl solution on the Data Sheet. Place the temperature probe in the solution and record the initial solution temperature on the Data Sheet. While still measuring the temperature of the solution, add 50.0 ml of 1.0 Molar NaOH (graduated cylinder or pipette) to the Styrofoam cup. [If a pipette is used, pipet the NaOH first into a glass beaker, and then pour the solution into a Styrofoam cup. The HCl and NaOH solutions need to have been thermally equilibrated at ambient room temperature. The solutions are generally prepared well in advance, so thermal equilibrium should not be a problem.] Gently stir the chemical reaction with the coffee stirrer provided. Keep the temperature probe in the center of the cup. This particular reaction is exothermic so the temperature should rise. After the temperature has leveled off, record the final temperature, $T_{final,soln}$, on the laboratory Data Sheet. Should a downward trend be observed in the plateau, take the maximum temperature measured as $T_{final,soln}$.

Weigh the Styrofoam cup and contents so the mass of solution, needed in Eqn. 11.2A, will be known. Record the mass on the Data Sheet. The mass of the solution = mass of Styrofoam cup + contents at the end of the experiment – mass of empty Styrofoam cup. Substitute the measured quantities into Eqn. 11.2A and calculate the enthalpy for the reaction of hydrochloric acid with sodium hydroxide. Use 4.184 J/(gram °C) for the specific heat of the solution. The value that you just calculated is for the formation of 0.0500 moles of water. The value is scaled up to the formation of one mole of water by dividing the measured enthalpy of reaction by the number of moles of water formed.

A theoretical value for the enthalpy of formation of water from hydrogen and hydroxide ions

$$H^+_{(aq)} + OH^-_{(aq)} \text{ ------> } H_2O_{(liq)}$$

can be calculated from

$$\Delta H_{reaction} = \Sigma \Delta H^o_{f,products} - \Sigma \Delta H^o_{f,reactants} \qquad (11A.3)$$

the standard molar enthalpies of formation of the reactants and products. Do not take time now to do this calculation. After the entire Experiment is over return to this step, and calculate the expected enthalpy of reaction for the formation of one mole of water. Values needed in this calculation are the standard molar enthalpies of formation: for liquid H_2O, $\Delta H^o_f = -285.83$ kJ/mole; for $H^+_{(aq)}$, $\Delta H^o_f = 0.00$ kJ/mole (by definition); and for $OH^-_{(aq)}$, $\Delta H^o_f = -230.0$ kJ/mole. How does your experimental value compare to the theoretical, expected value calculated from the standard molar enthalpies of formation? Calculate the relative percent deviation:

relative % deviation = 100 × (experiment value – correct value)/correct value

$$(11A.4)$$

between your measured value and the theoretical value.

ENTHALPY OF ACID-BASE REACTION – H₂SO₄ + 2 NaOH

To a first approximation one would expect that the enthalpy of reaction on a per mole of water formed basis would be the same for all acid-base reactions having the same net chemical reaction of:

$$H^+_{(aq)} + OH^-_{(aq)} \text{ ------> } H_2O_{(liq)}$$

Let us determine whether or not this is the case by measuring the enthalpy of reaction for

$$H_2SO_{4(aq)} + 2\,NaOH_{(aq)} \text{ -----> } 2\,H_2O + Na_2SO_{4(aq)}$$

mixing sulfuric acid and sodium hydroxide. The concentration of H_2SO_4 has been adjusted so that 0.0500 moles of water are formed whenever 50.0 ml of acid and 50.0 ml of base used.

Repeat the HCl-NaOH acid-base reaction, this time using 50.0 ml of 0.500 Molar H_2SO_4. Record the measured values on the laboratory Data Sheet under the section labeled "Enthalpy Determination for H_2SO_4 + 2 NaOH Reaction". Calculate the enthalpy of reaction by substituting the measured values into Eqn. 11.2. Divide the measured enthalpy by moles of liquid H_2O formed (0.0500) to put the enthalpy of reaction on a molar basis. How does the value for H_2SO_4 + NaOH reaction compare with the value for the HCl + NaOH reaction?

ENTHALPY OF SOLUTION – ANHYDROUS MAGNESIUM SULFATE (MgSO₄)

Anhydrous magnesium sulfate dissociates in water into

$$MgSO_{4(solid)} \ \text{------>} \ Mg^{2+}_{(aq)} + SO_4^{2-}_{(aq)}$$

hydrated Mg^{2+} and SO_4^{2-} ions. The reaction is exothermic, and the heat that is released when solid $MgSO_4$ dissociates will be absorbed the aqueous solution. The temperature of the solution will increase due to the absorption of heat.

To determine the enthalpy of solution of magnesium sulfate, you need to weigh an empty Styrofoam cup and record the mass on the laboratory Data Sheet in the appropriate space. Next add 100.0 ml of deionized water (graduated cylinder) and reweigh the cup + deionized water. Record the measured value so that the mass of water can be calculated by difference. Measure the initial temperature of the solution with the Pasco temperature probe. Next add about 5.00 to 6.00 grams (accurately weighed) of anhydrous magnesium sulfate to the Styroform cup while continuing to measure the temperature of the solution. Be sure that the mass of $MgSO_4$ is recorded on the Data Sheet as you will need this information in order to do the calculations. Stir with a plastic stirrer to mix and to facilitate dissolution. Keep the Pasco temperature probe in the center of the cup. Keep measuring the temperature of the solution until the dissolution is complete, and the temperature has stabilized. If a downward trend is observed, then use the maximum temperature reached for $T_{final,soln}$.

Calculate the enthalpy of solution of $MgSO_{4(solid)}$ by substituting the measured values into Eqn. 11A.2. [Note: Use 4.184 J/(gram °C) for the specific heat of the solution, and the total mass of the solution is the mass of water plus mass of magnesium sulfate.] Scale the calculated enthalpy of solution up to a per mole basis by dividing by the number of moles of $MgSO_4$ that dissolved. The molar mass of anhydrous magnesium sulfate (120.37 grams/mole) will be needed in scaling the measured enthalpy up to the per mole basis.

Once the entire experiment is over, come back to this step and calculate the expected theoretical value using the standard molar enthalpies of formation: for $MgSO_{4(solid)}$, $\Delta H^o_f = -1284.9$ kJ/mole; for $Mg^{2+}_{(aq)}$, $\Delta H^o_f = -466.85$ kJ/mole; and for $SO_4^{2-}_{(aq)}$, $\Delta H^o_f = -909.27$ kJ/mole. How does the measured value compare to the theoretical value?

[Note: Magnesium sulfate exists both as an anhydrous solid, $MgSO_{4(solid)}$, and has a heptahydrate, $MgSO_4 \cdot 7H_2O_{(solid)}$. In this experiment it is assumed that the anhydrous form is used. The anhydrous form is commercially available.]

ENTHALPY OF SOLUTION – AMMONIUM NITRATE (NH_4NO_3)

Repeat the magnesium sulfate procedure again with ammonium nitrate instead of magnesium sulfate, using 100.0 ml of water and 5.00 to 6.00 grams (accurately weighed) of ammonium nitrate. Be sure to record the actual mass of water and ammonium nitrate used. This particular reaction is endothermic, and the solution temperature will decrease as the solid dissolves. If an upward trend occurs in the plateau, use the minimum temperature reached for $T_{final,soln}$. The molar mass of ammonium nitrate (80.04 grams/mole) will be needed in scaling the measured enthalpy up to the per mole basis.

Once the entire experiment is over, come back to this step and calculate the expected theoretical value for the dissociation of ammonium nitrate

$$NH_4NO_{3(solid)} \text{ --------> } NH_4^+{}_{(aq)} + NO_3^-{}_{(aq)}$$

using the standard molar enthalpies of formation: for $NH_4NO_{3(solid)}$, $\Delta H^o{}_f = -365.56$ kJ/mole; for $NH_4^+{}_{(aq)}$, $\Delta H^o{}_f = -132.51$ kJ/mole; and for $NO_3^-{}_{(aq)}$, $\Delta H^o{}_f = -205.0$ kJ/mole. How does the measured value compare to the theoretical value?

ENTHALPY OF SOLUTION – SODIUM CHLORIDE (NaCl)

Repeat the magnesium sulfate procedure again with sodium chloride instead of magnesium sulfate, using 100.0 ml of water and 5.00 to 6.00 grams (accurately weighed) of sodium chloride. Be sure to record the actual mass of water and sodium chloride used. This particular reaction is slightly endothermic, and the solution temperature will decrease as the solid dissolves. If an upward trend occurs in the plateau, use the minimum temperature reached for $T_{final,soln}$. The molar mass of sodium chloride (58.44 grams/mole) will be needed in scaling the measured enthalpy up to the per mole basis.

Once the entire experiment is over, come back to this step and calculate the expected theoretical value for the dissociation of sodium chloride

$$NaCl_{(solid)} \text{ --------> } Na^+{}_{(aq)} + Cl^-{}_{(aq)}$$

using the standard molar enthalpies of formation: for $NaCl_{(solid)}$, $\Delta H^o{}_f = -411.15$ kJ/mole; for $Na^+{}_{(aq)}$, $\Delta H^o{}_f = -240.12$ kJ/mole; and for $Cl^-{}_{(aq)}$, $\Delta H^o{}_f = -167.2$ kJ/mole. How does the measured value compare to the theoretical value?

ENTHALPY OF SOLUTION – POTASSIUM NITRATE (KNO₃)

Repeat the magnesium sulfate procedure again with potassium nitrate instead of magnesium sulfate, using 100.0 ml of water and 5.00 to 6.00 grams (accurately weighed) of potassium nitrate. Be sure to record the actual mass of water and potassium nitrate used. This particular reaction is endothermic, and the solution temperature will decrease as the solid dissolves. If an upward trend occurs in the plateau, use the minimum temperature reached for $T_{final,soln}$. The molar mass of potassium nitrate (101.11 grams/mole) will be needed in scaling the measured enthalpy up to the per mole basis.

Once the entire experiment is over, come back to this step and calculate the expected theoretical value for the dissociation of potassium nitrate

$$KNO_{3(solid)} \dashrightarrow K^+_{(aq)} + NO_3^-_{(aq)}$$

using the standard molar enthalpies of formation: for $KNO_{3(ssolid)}$, $\Delta H^o_f = -492.70$ kJ/mole; for $K^+_{(aq)}$, $\Delta H^o_f = -252.4$ kJ/mole; and for $NO_3^-_{(aq)}$, $\Delta H^o_f = -205.0$ kJ/mole. How does the measured value compare to the theoretical value?

NOTE TO INSTRUCTOR – The laboratory manual was written with the intention that students would perform either Experiment 11A or Experiment 11B. The two experiments involve the same experimental methodology, and there is really no need to do both. Experiment 11A discusses the measured values more from the standpoint of standard enthalpies of formation, while Experiment 11B calculates the enthalpies of hydration using Hess's Law. Should one decide to assign Experiment 11B, the experiment should be short enough that students could also perform one of the Enthalpy of Acid – Base Reactions (HCl + NaOH, or H_2SO_4 + NaOH) from Experiment 11A in the three-hour laboratory period. The last part of the laboratory Data Sheet for Experiment 11B includes this provision.

DATA SHEET – EXPERIMENT 11A

Name: _____

Enthalpy Determination for HCl + NaOH Reaction

1. Mass of empty Styrofoam cup, g: _____

2. Mass of Styrofoam cup + HCl, g: _____

3. Initial temperature of HCl solution, °C: _____

4. Final temperature of solution after mixing, °C: _____

5. Change in temperature, °C: _____

6. Mass of Styrofoam cup with HCl and NaOH added, g: _____

7. Mass of solution, g: _____

8. ΔH of reaction, J: _____

9. ΔH of reaction, kJ/mole: _____

10. ΔH of reaction based on ΔH_f^o data, kJ/mole: _____

11. Percent deviation: _____

Enthalpy Determination for H₂SO₄ + 2 NaOH Reaction

12. Mass of empty Styrofoam cup, g: _____

13. Mass of Styrofoam cup + H_2SO_4, g: _____

14. Initial temperature of H_2SO_4 solution, °C: _____

15. Final temperature of solution after mixing, °C: _____

16. Change in temperature, °C: _____

17. Mass of Styrofoam cup with H_2SO_4 and NaOH added, g: _____

18. Mass of solution, g: _____

19. ΔH of reaction, J: _____

20. ΔH of reaction, kJ/mole: _____

21. ΔH of reaction based on ΔH_f^o data, kJ/mole: _____

22. Percent deviation: _____

Enthalpy of Solution – Magnesium Sulfate (MgSO₄)

23. Mass of empty Styrofoam cup, g: _____

24. Mass of Styrofoam cup + water, g: _____

25. Mass of $MgSO_4$ solid, g: _____

26. Mass of solution, g: _____

27. Initial temperature of water solution, °C: _____

28. Final temperature of solution after mixing, °C: _____

29. Change in temperature, °C: _____

30. ΔH of reaction, J: _____

31. ΔH of reaction, kJ/mole: _____

32. ΔH of reaction based on ΔH_f^o data, kJ/mole: _____

33. Percent deviation: _____

Enthalpy of Solution – Ammonium Nitrate (NH₄NO₃)

34. Mass of empty Styrofoam cup, g: _____

35. Mass of Styrofoam cup + water, g: _____

36. Mass of NH_4NO_3 solid, g: _____

37. Mass of solution, g: _____

38. Initial temperature of water solution, °C: _____

39. Final temperature of solution after mixing, °C: _____

40. Change in temperature, °C: _____

41. ΔH of reaction, J: _____

42. ΔH of reaction, kJ/mole: _____

43. ΔH of reaction based on ΔH_f^o data, kJ/mole: _____

44. Percent deviation: _____

Enthalpy of Solution – Sodium Chloride (NaCl)

45. Mass of empty Styrofoam cup, g: _____

46. Mass of Styrofoam cup + water, g: _____

47. Mass of NaCl solid, g: _____

48. Mass of solution, g: _____

49. Initial temperature of water solution, °C: _____

50. Final temperature of solution after mixing, °C: _____

51. Change in temperature, °C: _____

52. ΔH of reaction, J: _____

53. ΔH of reaction, kJ/mole: _____

54. ΔH of reaction based on ΔH_f^o data, kJ/mole: _____

55. Percent deviation: _____

Enthalpy of Solution – Potassium Nitrate (KNO₃)

56. Mass of empty Styrofoam cup, g: _____

57. Mass of Styrofoam cup + water, g: _____

58. Mass of KNO_3 solid, g: _____

59. Mass of solution, g: _____

60. Initial temperature of water solution, °C: _____

61. Final temperature of solution after mixing, °C: _____

62. Change in temperature, °C: _____

63. ΔH of reaction, J: _____

64. ΔH of reaction, kJ/mole: _____

65. ΔH of reaction based on ΔH_f^o data, kJ/mole: _____

66. Percent deviation: _____

EXPERIMENT 11B: DETERMINATION OF HEATS OF HYDRATION BASED ON HESS'S LAW

INTRODUCTION

It is not experimentally feasible to calorimetrically measure the reaction enthalpies of every known chemical reaction. Some chemical reactions are explosive, others involve potentially hazardous substances, and some reactions may take days for a noticeable change to be observed in the concentrations of the starting reactants. For such reactions, Hess's Law provides a very convenient means to obtain the enthalpy of reaction, $\Delta H_{reaction}$, for a desired chemical reaction by mathematically manipulating the measured reaction enthalpies (heats) for two or more other chemical reactions.

In today's laboratory experiment you will determine the enthalpy of hydration of magnesium sulfate heptahydrate, $MgSO_4 \cdot 7H_2O_{(solid)}$:

$$MgSO_{4(solid)} + 7\,H_2O \quad \text{------->} \quad MgSO_4 \cdot 7H_2O_{(solid)}$$

and sodium sulfate decahydrate, $Na_2SO_4 \cdot 10H_2O_{(solid)}$:

$$Na_2SO_{4(solid)} + 10\,H_2O \quad \text{------->} \quad Na_2SO_4 \cdot 10H_2O_{(solid)}$$

by measuring the enthalpies of solution of both the anhydrous and hydrated salt in water. For magnesium sulfate, the enthalpies of solution would correspond to:

$$MgSO_{4(solid)} \quad \text{------->} \quad Mg^{2+}_{(aq)} + SO_4^{2-}_{(aq)}$$

$$MgSO_4 \cdot 7H_2O_{(solid)} \quad \text{-------->} \quad Mg^{2+}_{(aq)} + SO_4^{2-}_{(aq)} + 7\,H_2O$$

Subtraction of the second solution equation from the first gives the chemical reaction describing the hydration of anhydrous magnesium sulfate to form the heptahydrate salt. The molar solution enthalpies (per mole of salt) are similarly combined, to give

molar enthalpy of hydration of $MgSO_{4(solid)}$ = molar enthalpy of solution of

$MgSO_{4(solid)}$ - molar enthalpy of solution of $MgSO_4 \cdot 7H_2O_{(solid)}$

(11B.1)

the molar enthalpy of hydration. In many respects Hess's law is nothing more than mathematically manipulating a set of chemical equations in order to derive the desired reaction. Hess's law is another way of stating the conversation of energy. In an ordinary chemical reaction energy can neither be created nor destroyed. Hess's law works even if the overall reaction does not actually occur in the manner the separate chemical reactions were combined. Remember enthalpy is a thermodynamic state function, and the change in enthalpy is independent of path.

The four specific reactions that are to be studied today take place in aqueous solution. The Styrofoam cup calorimeter will again be used. The calorimeter is not sealed, so the reactions occur under the essentially constant pressure of the atmosphere. Our experience has shown that Styrofoam cups are sufficiently insulated, and that transfer of heat between the solution and its outside surrounds is not a major concern. Styrofoam has a very low thermal conductivity and specific heat. For a good adiabatic calorimeter, the heat gained (or lost) by the solution must be:

$$\text{heat of reaction} = - \text{heat lost or gained by solution} \tag{11B.2}$$

equal to, but opposite in sign, as the heat evolved (or gained) by the reaction. The right-hand side of Eqn. 11B.2 is calculated from

$$\text{heat of reaction} = - (\text{SpHt of solution}) \times (\text{mass of solution}) \times (T_{\text{final,soln}} - T_{\text{initial,soln}})$$
$$\tag{11B.3}$$

the mass of the solution, its specific heat, and the temperature change. For dilute aqueous solutions, the specific heat of the solution will be approximately the same as that of water, 4.184 J/(gram °C).

ENTHALPY OF SOLUTION – ANHYDROUS MAGNESIUM SULFATE ($MgSO_4$)

Anhydrous magnesium sulfate dissociates in water into

$$MgSO_{4(solid)} \ \ ------> \ \ Mg^{2+}_{(aq)} + SO_4^{2-}_{(aq)}$$

hydrated Mg^{2+} and SO_4^{2-} ions. The reaction is exothermic, and the heat that is released when solid $MgSO_4$ dissociates will be absorbed the aqueous solution. The temperature of the solution will increase due to the absorption of heat. A Pasco temperature probe is used to measure the initial and final solution temperature. Set the Pasco system up to measure temperature and calibrate the temperature probe. For this experiment you should be able to use an ice-water bath for the lower reference temperature (0 °C) and warm water (in the 40 – 50 °C temperature range) for the upper reference temperature. The instructions for calibrating the temperature probe, and for performing time versus temperature measurements, are found in Experiment 8.

To determine the enthalpy of solution of anhydrous magnesium sulfate, you need to weigh an empty Styrofoam cup and record the mass on the laboratory Data Sheet in the appropriate space. Next add 100.0 ml of deionized water (graduated cylinder) and reweigh the cup + deionized water. Record the measured value so that the mass of water can be calculated by difference. Measure the initial temperature of the solution with the Pasco temperature probe. Next add about 5.00 to 6.00 grams (accurately weighed) of anhydrous magnesium sulfate to the Styroform cup while continuing to measure the temperature of the solution. Be sure that the mass of $MgSO_4$ is recorded on the Data Sheet as you will need this information in order to do the calculations. Stir with a plastic stirrer to mix and to facilitate dissolution. Keep the Pasco temperature probe in the center of the cup. Keep measuring the temperature of the solution until the dissolution is complete, and the temperature has stabilized. If a downward trend is observed, then use the maximum temperature reached for $T_{final,soln}$.

Calculate the enthalpy of solution of $MgSO_{4(solid)}$ by substituting the measured values into Eqn. 11B.3. [Note: Use 4.184 J/(gram °C) for the specific heat of the solution, and the total mass of the solution is the mass of water plus mass of magnesium sulfate.] Scale the calculated enthalpy of solution up to a per mole basis by dividing by the number of moles of $MgSO_4$ that dissolved. The molar mass of anhydrous magnesium sulfate (120.37 grams/mole) will be needed in scaling the measured enthalpy up to the per mole basis.

ENTHALPY OF SOLUTION – MAGNESIUM SULFATE HEPTAHYDRATE ($MgSO_4 \cdot 7H_2O$)

Magnesium sulfate heptahydrate is commercially available, and in today's experiment it will be assumed that every molecule does indeed have seven waters of hydration. One could experimentally determine whether or not this is true by actually measuring the hydration number by heating the solid over a Bunsen burner to remove the water, as was done in Experiment 3.

Repeat the magnesium sulfate procedure again with magnesium sulfate heptahydrate instead of magnesium sulfate, using 100.0 ml of water and 5.00 to 6.00 grams (accurately weighed) of magnesium sulfate heptahydrate. Be sure to record the actual mass of water and magnesium sulfate heptahydrate used as this value will be needed to scale the measured enthalpy of solution up to a per mole basis. The dissolution should be slightly endothermic. If an upward trend occurs in the plateau, use the minimum temperature reached for $T_{final,soln}$. Be sure to record the initial and final temperatures on the laboratory Data Sheet. Calculate the enthalpy of solution of $MgSO_4 \cdot 7H_2O_{(solid)}$ by substituting the measured values into Eqn. 11B.3. [Note: Use 4.184 J/(gram °C) for the specific heat of the solution, and the total mass of the solution is the mass of water plus mass of magnesium sulfate heptahydrate.] Scale the calculated enthalpy of solution up to a per mole basis by dividing by the number of moles of $MgSO_4 \cdot 7H_2O_{(solid)}$ that dissolved. The molar mass of magnesium sulfate heptahydrate (246.48 grams/mole) will be needed in scaling the measured enthalpy up to the per mole basis.

The molar enthalpy of hydration of anhydrous magnesium sulfate is calculated by substituting the molar enthalpies of solution of the anhydrous and hydrated salt into Eqn. 11B.1. Literature values for the enthalpies of solution of $MgSO_4$ and $MgSO_4 \cdot 7H_2O$ are

$\Delta H = -84.85$ kJ/mole and $\Delta H = 15.90$ kJ/mole, respectively. How does your measured value for the enthalpy of hydration compare to the value calculated using the literature enthalpy of solution data?

ENTHALPY OF SOLUTION – ANHYDROUS SODIUM SULFATE (Na_2SO_4)

Anhydrous sodium sulfate dissociates in water into

$$Na_2SO_{4(solid)} \;\text{------>}\; 2\,Na^+_{(aq)} + SO_4^{2-}{}_{(aq)}$$

hydrated Na^+ and SO_4^{2-} ions. This reaction is very, very slightly exothermic. You may not even observe much in the way of a temperature change. Record whatever temperature change you observed, and the enthalpy of solution of $Na_2SO_{4(solid)}$ by substituting the measured values into Eqn. 11B.3. [Note: Use 4.184 J/(gram °C) for the specific heat of the solution, and the total mass of the solution is the mass of water plus mass of anhydrous sodium sulfate.] Scale the calculated enthalpy of solution up to a per mole basis by dividing by the number of moles of $Na_2SO_{4(solid)}$ that dissolved. The molar mass of anhydrous sodium sulfate (142.04 grams/mole) will be needed in scaling the measured enthalpy up to the per mole basis.

ENTHALPY OF SOLUTION – SODIUM SULFATE DECAHYDRATE ($Na_2SO_4 \cdot 10H_2O$)

Repeat the magnesium sulfate procedure again with sodium sulfate decahydrate instead of magnesium sulfate, using 100.0 ml of water and 5.00 to 6.00 grams (accurately weighed) of sodium sulfate decahydrate. Be sure to record the actual mass of water and sodium sulfate decahydrate used as this value will be needed to scale the measured enthalpy of solution up to a per mole basis. The dissolution should be endothermic. If an upward trend occurs in the plateau, use the minimum temperature reached for $T_{final,soln}$. Be sure to record the initial and final temperatures on the laboratory Data Sheet. Calculate the enthalpy of solution of $Na_2SO_4 \cdot 10H_2O_{(solid)}$ by substituting the measured values into Eqn. 11B.3. [Note: Use 4.184 J/(gram °C) for the specific heat of the solution, and the total mass of the solution is the mass of water plus mass of sodium sulfate decahydrate.] Scale the calculated enthalpy of solution up to a per mole basis by dividing by the number of moles of $Na_2SO_4 \cdot 10H_2O_{(solid)}$ that dissolved. The molar mass of sodium sulfate decahydrate (322.19 grams/mole) will be needed in scaling the measured enthalpy up to the per mole basis.

The molar enthalpy of hydration of anhydrous sodium sulfate is calculated by substituting the molar enthalpies of solution of the anhydrous and hydrated salt into Eqn.

11B.1. Literature values for the enthalpies of solution of Na_2SO_4 and $Na_2SO_4 \cdot 10H_2O$ are $\Delta H = -1.92$ kJ/mole and $\Delta H = 78.49$ kJ/mole, respectively. How does your measured value for the enthalpy of hydration compare to the value calculated using the literaure enthalpy of solution data?

> **NOTE TO INSTRUCTOR** – The laboratory manual was written with the intention that students would perform either Experiment 11A or Experiment 11B. The two experiments involve the same experimental methodology, and there is really no need to do both. Experiment 11A discusses the measured values more from the standpoint of standard enthalpies of formation, while Experiment 11B calculates the enthalpies of hydration using Hess's Law. Should one decide to assign Experiment 11B, the experiment should be short enough that students could also perform one of the Enthalpy of Acid – Base Reactions (HCl + NaOH, or H_2SO_4 + NaOH) from Experiment 11A in the three-hour laboratory period. The last part of the laboratory Data Sheet for Experiment 11B includes this provision.

DATA SHEET – EXPERIMENT 11B

Epsom salt

Name: _____

Enthalpy of Solution – Anhydrous Magnesium Sulfate ($MgSO_4$)

1. Mass of empty Styrofoam cup, g: _____

2. Mass of Styrofoam cup + water, g: _____

3. Mass of $MgSO_4$ solid, g: _____

4. Mass of solution, g: _____

5. Initial temperature of water solution, °C: _____

6. Final temperature of solution after mixing, °C: _____

7. Change in temperature, °C: _____

8. ΔH of reaction, J: _____

9. ΔH of reaction, kJ/mole: _____

Enthalpy of Solution – Magnesium Sulfate Heptahydrate ($MgSO_4 \cdot 7H_2O$)

10. Mass of empty Styrofoam cup, g: _____

11. Mass of Styrofoam cup + water, g: _____

12. Mass of $MgSO_4 \cdot 7H_2O$ solid, g: _____

13. Mass of solution, g: _____

14. Initial temperature of water solution, °C: _____

15. Final temperature of solution after mixing, °C: _____

16. Change in temperature, °C: _____

17. ΔH of reaction, J: _____

18. ΔH of reaction, kJ/mole: _____

Enthalpy of Hydration of Anhydrous Magnesium Sulfate

19. ΔH of hydration, kJ/mole: _____

Enthalpy of Solution – Anhydrous Sodium Sulfate (Na$_2$SO$_4$)

20. Mass of empty Styrofoam cup, g: _____

21. Mass of Styrofoam cup + water, g: _____

22. Mass of Na$_2$SO$_4$ solid, g: _____

23. Mass of solution, g: _____

24. Initial temperature of water solution, ºC: _____

25. Final temperature of solution after mixing, ºC: _____

26. Change in temperature, ºC: _____

27. ΔH of reaction, J: _____

28. ΔH of reaction, kJ/mole: _____

Enthalpy of Solution – Sodium Sulfate Decahydrate (Na$_2$SO$_4$·10H$_2$O)

29. Mass of empty Styrofoam cup, g: _____

30. Mass of Styrofoam cup + water, g: _____

31. Mass of Na$_2$SO$_4$·10H$_2$O solid, g: _____

32. Mass of solution, g: _____

33. Initial temperature of water solution, ºC: _____

34. Final temperature of solution after mixing, ºC: _____

35. Change in temperature, ºC: _____

36. ΔH of reaction, J: _____

37. ΔH of reaction, kJ/mole: _____

Enthalpy of Hydration of Anhydrous Sodium Sulfate

38. ΔH of hydration, kJ/mole: _____

Enthalpy Determination for Acid + Base Neutralization Reaction

39. Name of acid used: _____

40. Name of base used: _____

41. Mass of empty Styrofoam cup, g: _____

42. Mass of Styrofoam cup + acid, g: _____

43. Initial temperature of acid solution, °C: _____

44. Final temperature of solution after mixing, °C: _____

45. Change in temperature, °C: _____

46. Mass of Styrofoam cup with acid and base added, g: _____

47. Mass of solution, g: _____

48. ΔH of reaction, J: _____

49. ΔH of reaction, kJ/mole: _____

EXPERIMENT 12: INTRODUCTION TO ORGANIC CHEMISTRY – SYNTHESIS OF ASPIRIN OF AND ORGANIC ESTERS

INTRODUCTION

Organic chemistry is the study of the properties and reactions of carbon-containing compounds. Although a few compounds containing carbon, such as carbides and carbonates are considered to be inorganic substances, the vast majority are organic compounds that typically contain chains or rings of carbon atoms. Numerous organic compounds occur naturally in living systems, and many more are produced through commercial manufacturing processes. Commercial products that you might already be familiar with include synthetic fibers, plastics, polymers, artificial sweeteners and a whole host of pharmaceutical drug molecules. In addition, the energy that we rely on so heavily for our transportation and heating needs is based on the organic materials found in coal and petroleum.

Inorganic chemical reactions are learned and remembered in terms of specific compounds. That is one mole of nitrogen gas reacts with three moles of hydrogen gas to form two moles of ammonia gas:

$$1 \text{ mole of } N_{2(gas)} + 3 \text{ moles of } H_{2(gas)} \quad \text{----------} > \quad 2 \text{ moles of } NH_{3(gas)}$$

One mole of nitrogen gas reacts with two moles of hydrogen gas to form one mole of hydrazine:

$$1 \text{ mole of } N_{2(gas)} + 2 \text{ moles of } H_{2(gas)} \quad \text{----------} > \quad 1 \text{ mole of } H_2NNH_2$$

Two moles of hydrogen gas react with 1 mole of oxygen gas to form two moles of water:

$$2 \text{ moles of } H_{2(gas)} + 1 \text{ mole of } O_{2(gas)} \text{ ----------} > 2 \text{ moles of } H_2O$$

One mole of hydrogen gas reacts with 1 mole of oxygen gas to form one mole of hydrogen peroxide:

$$1 \text{ moles of } H_{2(gas)} + 1 \text{ mole of } O_{2(gas)} \text{ ----------} > 1 \text{ mole of } H_2O_2$$

Compare this to the way that organic chemical reactions are studied and remembered using functional groups. A molecule containing functional group A reacts with a molecule containing functional group B to form a molecule having functional group C:

$$\text{Functional group A} + \text{Functional group B} \quad \text{----------} > \quad \text{Functional group C}$$

For an organic chemical reaction there would be a set of reaction conditions (such as perhaps heat, a specific solvent, presence of acid, presence of base, *etc.*) that would be required in order for the chemical transformation to occur.

What is a functional group? A functional group is a specific arrangement of atoms in the molecule that gives the molecule its physical and chemical properties. For example, alcohols are organic molecules that the "-OH" group in the molecule as evidence in the following molecular structures:

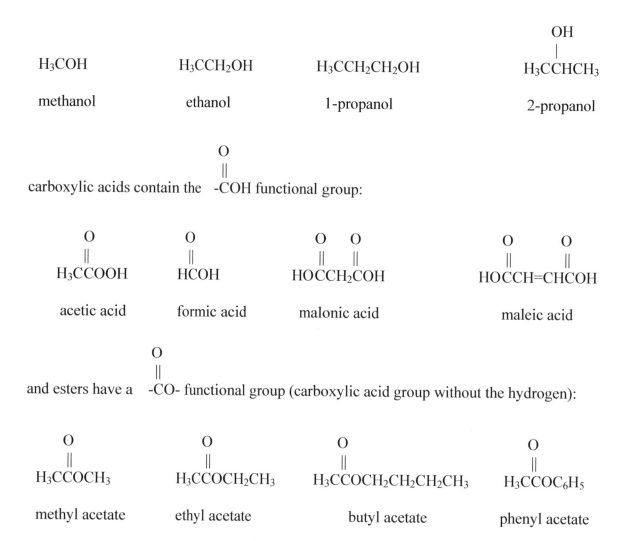

H₃COH

methanol

H₃CCH₂OH

ethanol

H₃CCH₂CH₂OH

1-propanol

$$\overset{\displaystyle OH}{\underset{\displaystyle}{|}}$$
H₃CCHCH₃

2-propanol

carboxylic acids contain the $-\overset{O}{\overset{\|}{C}}OH$ functional group:

H₃CCOOH

acetic acid

HCOH

formic acid

HOCCH₂COH

malonic acid

HOCCH=CHCOH

maleic acid

and esters have a $-\overset{O}{\overset{\|}{C}}O-$ functional group (carboxylic acid group without the hydrogen):

H₃CCOCH₃

methyl acetate

H₃CCOCH₂CH₃

ethyl acetate

H₃CCOCH₂CH₂CH₂CH₃

butyl acetate

H₃CCOC₆H₅

phenyl acetate

with both ends of the ester group connected to carbon atoms. Most esters have pleasant odors, and are responsible for the fragrances of many flowers and for the pleasant taste of ripened fruit. For example, bananas contain pentyl acetate and oranges contain octyl acetate. Table 12.1 lists the names of several esters found in common fruits or vegetables and/or what the fragrance resembles.

Table 12.1. Selected organic esters found in common fruits and flowers

Name of Organic Ester	Fruit or Flower
Ethyl formate	rum
Isobutyl formate	raspberry
Ethyl acetate	nail polish remover
Propyl acetate	pear
Butyl acetate	banana
Isopentyl acetate	banana
Pentyl acetate	banana
Octyl acetate	orange
Benzyl acetate	peach, jasmine
Menthyl acetate	peppermint
Linalyl acetate	bergamot
Ethyl propionate	butterscotch
Isobutyl propionate	rum
Pentyl propionate	apricot
Methyl butanoate	apple, pineapple
Ethyl butanoate	apple, pineapple
Butyl butanonate	pineapple
Pentyl butanoate	apricot
Benzyl butanoate	rose
Ethyl isobutanoate	apple
Isopentyl pentanoate	apple
Isopentyl isopenanoate	apple
Ethyl heptanoate	grape
Ethyl nonanoate	cognac
Ethyl decanoate	grape
Methyl benzoate	marzipan
Methyl salicylate	oil of wintergreen
Hexyl salicylate	azalea
Methyl anthranilate	grape
Phenyl ethyl tiglate	rose
Ethyl cinnamate	cinnamon

The chemical reaction that will be studied today involves:

Alcohol + Carboxylic Acid ---------> Ester + Water

the esterification of a carboxylic acid and an alcohol to form an organic ester and water. The reaction requires heat and a small amount of strong acid, such as sulfuric acid. This is one of the classic organic chemical reactions that one studies in the second year organic chemistry course. You will synthesize aspirin (chemical name is acetylsalicylic acid) and determine the percent yield. On the much smaller "test-tube" scale you will also make several of the compounds given in Table 12.1. Salicylic acid is a naturally occurring substance known to have both analgesic and anti-inflammatory properties; however, the compound itself severely irritates the intestinal lining of the stomach. The form of salicylic acid that is produced for human consumption is the acetyl ester (better known as aspirin), which breaks down in the human body into salicylic acid. In the pharmaceutical industry, such compounds that break down into the "active" drug in the body are called prodrug molecules.

The aspirin percent yield calculation is based on the limiting reagent, which is salicylic acid. Acetic anhydride is added in excess. Although the chemical reaction is written in terms of acetic anhydride (an initial starting reactant), the acetic acid that is produced from the reaction undoubtedly also reacts with the dissolved salicylic acid. Based on the reaction stoichiometry from the balanced chemical equation, one mole of salicylic acid should produce one mole of aspirin. The percent theoretical yield is:

% theoretical yield = 100 × (actual yield of aspirin/expected yield of aspirin)

$$(12.1)$$

moles of aspirin expected = moles of salicylic acid used $\qquad (12.2)$

$$\frac{grams\ of\ aspirin\ expected}{molar\ mass\ of\ aspirin} = \frac{grams\ of\ salicylic\ acid\ use}{molar\ mass\ of\ salicylic\ acid} \qquad (12.3)$$

calculated from the grams of aspirin actually produced and the grams of salicylic acid that one started with. The molar masses of salicylic acid and acetylsalicylic acid are 138.13 grams/mole and 180.17 grams/mole, respectively. No chemical reaction is 100 % efficient. Based on past experience a good yield for this synthesis would be 85 % to 95 %.

Chemists perform synthetic experiments on the "laboratory bench-top" scale. No attempt is really made to produce the compound at a net profit. To scale today's reaction up to an industrial manufacturing scale, a chemist would need to carefully examine how changes in the reaction conditions and reagent concentrations affect the overall percent yield and reaction time. Such a study would involve chemical equilibrium and chemical kinetic considerations. Next semester you will perform experiments related to both topics. On a manufacturing scale a one or two percent increase in the percent yield, which might seem as only a small increase for a laboratory bench-top synthesis, would translate to a considerable profit for the company. Considerable time and effort is spent in the chemical industry trying to increase product yields by improving the overall efficiency of chemical reactions.

EXPERIMENTAL PROCEDURE FOR ASPIRIN SYNTHESIS

Weight a 125-ml Erlenmeyer flask on a top loading electronic balance and record the mass on the laboratory Data Sheet in the section titled "Synthesis of Aspirin". Next add about 3 grams of salicylic acid to the flask and reweigh. Record the mass of the flask + salicylic acid on the Data Sheet. The mass of salicylic acid is obtained by difference. Salicylic acid is the limiting reagent in the aspirin synthesis, and its mass will be needed to perform the percent yield computation.

CAUTION – SALICYLIC ACID IS A SKIN IRRITANT. IF ANY GETS ON YOUR HANDS WASH IT OFF LOTS OF WATER.

Go to the hood and carefully add 6 ml of acetic anhydride to the flask. The acetic anhydride should just cover the salicylic acid.

CAUTION – ACETIC ANHYDRIDE IS BOTH A SKIN AND EYE IRRITANT.

Next carefully add 6 ml of concentrated sulfuric acid (Note to Instructor - 85 % phosphoric acid can be used instead of sulfuric acid) to the mixture.

CAUTION – CONCENTRATED SULFURIC ACID IS CORROSIVE AND CAN CAUSE SEVERE BURNS IF LEFT IN CONTACT WITH THE SKIN. IF YOU SPLASH ANY ON YOU, WASH IT OFF WITH PLENTY OF COOL WATER FOR SEVERAL MINUTES AND INFORM YOUR TEACHING ASSISTANT.

Cover the top of the flask with a piece of aluminum foil. Clamp the flask in a water bath for about five minutes. Make sure that the surface of the water bath is above the surface of the liquid in the Erlenmeyer flask so that the entire reaction mixture is heated. The temperature of the water bath should be at least 80 °C. If a hot water laboratory bath is not available, you can set up your own using a large beaker and Bunsen burner (or hotplate) to heat the water to at least 80 °C. After the reaction mixture has heated for 5 minutes, remove the flask from the hot water bath. Allow the reaction mixture to cool for 5 minutes, and then very cautiously add 20 ml of ice water to decompose any unreacted acetic anhydride. Now place the 125-ml Erlenmeyer flask in an ice-water bath to cool. Crystals of aspirin (acetylsalicylic acid) should start to form. If crystallization of the product does not occur during the cooling process, you might try scratching the inside of the test tube with a glass pipette.

When you no longer visually see any more crystals forming, it will be time to collect the product by vacuum filtration. This is the same experimental procedure used in

Experiment 5 to collect the alum. Clamp a filtering flask with side-arm to a ring stand, and attach a rubber tube between the side-arm and an aspirator on the water faucet. When the water is turned on, the flow past the aspirator creates a sunction. Place the Büchner filter funnel into the mouth of the side-arm filter, turn on the water faucet to create suction, and moisten the filter paper with a little warm from a wash bottle to hold it in place. Carefully pour your reaction mixture into the funnel.

After you have filtered the reaction mixture, add about 10 ml of ice to the 125-ml Erlenmeyer reaction flask, swirl around to pick up any residual solid that might be clinging to the sides of the reaction flask, and pour this into the funnel as well. Rinse the crystals with another 10 ml portion of ice-cold water to remove any impurities that might be adsorbed onto the surface of the crystals. The rinse will not remove impurities that might be trapped inside the crystals. While the crystals are air drying, weigh a clean, dry 100-ml or 150-ml beaker. Record the mass of the empty beaker on the Data Sheet. When the crystals appear dry transfer them to the beaker that you just weighed, reweigh the beaker with crystals, and record the mass of beaker + aspirin on the Data Sheet. The mass of the aspirin (actual yield in Eqn. 12.1) that was produced is calculated by difference.

The aspirin was crystallized fairly fast from solution, and it is undoubtedly very impure.

CAUTION – THE ASPIRIN IS IMPURE AND IS NOT INTENDED FOR CONSUMPTION. UNDER NO CIRCUMSTANCES SHOULD YOU INGEST YOUR ASPIRIN. IT WILL MAKE YOU SERIOUSLY ILL.

The aspirin could be purified by dissolving it in a small amount of a suitable solvent (G. A. Mirafzal and J. M. Summer, *J. Chem. Educ.*, **77**, 356-357 (2000) used toluene as the recrystallization solvent), and then allowing the solution to cool very, very slowly. Very slow cooling can be accomplished by wrapping the container in aluminum foil. Impurities are less likely to be trapped in slow growing crystals. Rather than spend time today doing another recrystallization, the remainder of the laboratory time will be spent on synthesizing several additional esters on the test-tube scale. For those of you planning to take second-year organic chemistry, do not worry, you will perform many recrystallization purifications in next year's chemistry laboratory course.

SYNTHESIS OF VARIOUS ESTERS ON THE TEST-TUBE SCALE

Table 12.2 gives the starting reagents and suggested reaction for preparing several esters that exhibit fruity or flowery odors. These esters were selected to give water-soluble byproducts at the end of the reaction, which should facilitate the separation of the ester layer as the top phase. You will not have time to synthesize every compound in the table, but if possible try to synthesize at least three or four of the compounds so that you can see for yourself that "yes, organic esters do exhibit fruity and/or flowery fragrances". The molar structures of the alcohol and carboxylic acid starting reactants are given in Table 12.3. You will need the molecular structures in order to answer some of the questions on the laboratory report form.

Add a small amount of each alcohol and carboxylic acid pair that your want to study into a test tube. Each test tube should have one alcohol and one carboxylic acid. Cautiously add 3 drops of concentrated sulfuric acid. **[Note the safety cautionary warning in the preceding section.** Sulfuric acid acts as catalyst to speed up the reaction.] Gently swirl the test tube until all of the chemicals are dissolved. Place the test tube in a hot water bath (heated with a Bunsen burner or hot plate), which is held at a temperature between 90 – 100 °C. Heat the reaction mixture for about 10 minutes. Watch the reaction mixture carefully. The lower-boiling alcohols (methyl alcohol, ethyl alcohol, propyl alcohol and isopropyl alcohol) may begin to boil, in which case you will need to lower the water bath temperature a few degrees (down to 80 °C or so). Alternatively, you could occasionally remove the test tube from the hot water bath with a test tube holder. After the reaction has heated for 10 minutes, remove the test tube with a test tube holder, and allow to cool in another beaker full of water. Once the reaction mixture has cooled down to ambient room temperature, add deionized water to fill the test tube up to about the halfway mark. Cautiously add a spatula-tip full of sodium bicarbonate ($NaHCO_3$). The sodium bicarbonate will neutralize any excess, unreacted carboxylic acid that might be present. If there is excess acid, the solution will start bubbling/fizzing when the sodium bicarbonate is added. Do not add the sodium bicarbonate too fast as you want to prevent the contents from fizzing over. Stir the solution carefully, and be careful of "overfizzing". At this point you should be able to observe a thin organic layer forming at the top of the mixture. Dip a Q-tip or a small piece of rolled up paper towel into the upper organic phase layer and remove. Very cautiously smell and record the smell on the laboratory Data Sheet in the section titled "Test-tube Scale Synthesis of Esters".

Table 12.2. Recommended List of Possible Esters to Prepare and the Respective Alcohol and Carboxylic Acid Starting Reagents

Alcohol	Amount	Acid	Amount	Product
Ethyl alcohol	3 ml	acetic acid	1 ml	ethyl acetate
Isopentyl alcohol	1 ml	acetic acid	3 ml	isopentyl acetate
Benzyl alcohol	0.5 ml	acetic acid	3 ml	benzyl acetate
Octyl alcohol	0.5 ml	acetic acid	3 ml	octyl acetate
Butyl alcohol	0.5 ml	acetic acid	3 ml	butyl acetate
Isopentyl alcohol	1 ml	propionic acid	3 ml	isopentyl propionate
Methyl alcohol	3 ml	benzoic acid	0.5 g	methyl benzoate
Methyl alcohol	3 ml	salicylic acid	0.5 g	methyl salicylate
Ethyl alcohol	3 ml	octanoic acid	0.5 ml	ethyl octanoate
Isopentyl alcohol	1 ml	pentanoic acid	0.5 ml	isopentyl pentanoate
Ethyl alcohol	1 ml	*trans*-cinnamic acid	2 ml	ethyl cinnamate
Ethyl alcohol	1.5 ml	propionic acid	2 ml	ethyl propionate

Table 12.3. Molecular Structures of Starting Alcohol and Carboxylic Acid Reactants

Compound Name	Molecular Structure
Alcohol:	
Methyl alcohol	CH_3OH
Ethyl alcohol	CH_2CH_2OH
Propyl alcohol	$CH_2CH_2CH_2OH$
Isopropyl alcohol	$(CH_3)_2CHOH$
Butyl alcohol	$CH_2CH_2CH_2CH_2OH$
Isobutyl alcohol	$CH_3CH(CH_3)CH_2OH$
tertiary-Butyl alcohol	$(CH_3)_3COH$
Pentyl alcohol	$CH_2CH_2CH_2CH_2CH_2OH$
Isopentyl alcohol	$(CH_3)_2CHCH_2CH_2OH$
Hexyl alcohol	$CH_2CH_2CH_2CH_2CH_2CH_2OH$
Heptyl alcohol	$CH_2CH_2CH_2CH_2CH_2CH_2CH_2OH$
Octyl alcohol	$CH_2CH_2CH_2CH_2CH_2CH_2CH_2CH_2OH$
Phenol	C_6H_5OH
Benzyl alcohol	$C_6H_5CH_2OH$
Carboxylic Acid:	
Formic acid	$HCOOH$
Acetic acid	CH_3COOH
Propionic acid	CH_3CH_2COOH
Butanoic acid	$CH_3CH_2CH_2COOH$
Pentanoic acid	$CH_3CH_2CH_2CH_2COOH$
Hexanoic acid	$CH_3CH_2CH_2CH_2CH_2COOH$
Octanoic acid	$CH_3CH_2CH_2CH_2CH_2CH_2CH_2COOH$
Benzoic acid	C_6H_5COOH
Malonic acid	$HOOCCH_2COOH$
Adipic acid	$HOOCCH_2CH_2CH_2CH_2COOH$
Cinnamic acid	$C_6H_5CH=CHCOOH$
Terephthalic acid	$HOOCC_6H_4COOH$

DATA SHEET – EXPERIMENT 12

Name: _____

Synthesis of Aspirin

1. Mass of empty flask, g: _____

2. Mass of flask + salicylic acid, g: _____

3. Mass of salicylic acid, g: _____

4. Number of moles of salicylic acid used: _____

5. Number of moles of aspirin expected: _____

6. Expected yield of aspirin in grams, g: _____

7. Mass of empty beaker, g: _____

8. Mass of beaker + aspirin, g: _____

9. Mass of aspirin, g: _____

10. Percent yield of aspirin: _____

11. Discussion of reasons why the percent yield might be less than 100 %:

Test-Tube Scale Synthesis of Esters

Synthesis of First Ester:

12. Name of alcohol used: _____

13. Name of carboxylic acid used: _____

14. Describe odor of ester synthesized: _____

15. Write a chemical reaction in terms of structural formulas that describes the ester that you synthesized. (See Table 12.3 for molecular structures of the starting reactants. You should be able to figure out the molecular structure for the product based on an understanding of functional group reaction.)

Synthesis of Second Ester:

16. Name of alcohol used: _____

17. Name of carboxylic acid used: _____

18. Describe odor of ester synthesized: _____

19. Write a chemical reaction in terms of structural formulas that describes the ester that you synthesized. (See Table 12.3 for molecular structures of the starting reactants. You should be able to figure out the molecular structure for the product based on an understanding of functional group reaction.)

Synthesis of Third Ester:

20. Name of alcohol used: _____

21. Name of carboxylic acid used: _____

22. Describe odor of ester synthesized: _____

23. Write a chemical reaction in terms of structural formulas that describes the ester that you synthesized. (See Table 12.3 for molecular structures of the starting reactants. You should be able to figure out the molecular structure for the product based on an understanding of functional group reaction.)

Synthesis of Fourth Ester:

24. Name of alcohol used: _____

25. Name of carboxylic acid used: _____

26. Describe odor of ester synthesized: _____

27. Write a chemical reaction in terms of structural formulas that describes the ester that you synthesized. (See Table 12.3 for molecular structures of the starting reactants. You should be able to figure out the molecular structure for the product based on an understanding of functional group reaction.)

Post Laboratory Exercise

In the design of new synthetic methods, organic chemists often work backwards from the target molecule in order to select the appropriate starting reactants. For the five chemical reactions below write the molecular structures of the starting alcohol and carboxylic acid needed to make the desired organic ester. (Hint: See Table 12.3 for the molecular structures of alcohols and carboxylic acids.)

28. _____ + _____
          alcohol   acid   --------> $CH_3COOC(CH_3)_3$

29. _____ + _____
          alcohol   acid   --------> $CH_3CH_2COOC_6H_5$

30. _____ + _____
          alcohol   acid   --------> $HCOOCH(CH_3)_2$

31. _____ + _____
          alcohol   acid   --------> $CH_3OOCC_6H_4COOCH_3$

32. _____ + _____
          alcohol   acid   -------> $(CH_3)_2CHOOCCH_2CH_2CH_3$

EXPERIMENT 13: CHEMICAL KINETICS I – DETERMINATION OF THE ORDER OF REACTION AND RATE CONSTANT BASED ON DIFFERENTIAL RATE FORM EXPRESSION

INTRODUCTION

Chemical kinetics is the study of reaction rates, how reaction rates change under varying conditions, and what series of molecular events or steps occur during the overall reaction. In today's laboratory experimental you will examine how the rate of a chemical reaction, lets see for chemicals A and B reacting:

Chemical A + Chemical B ------> Products

changes when the concentration of one or both reactants is increased. You may already have some ideas in this regard. For two molecules to react, they must come in close proximity to each other, and have sufficient energy upon collision to get over the initial energy of activation barrier. From a simple statistical point-of-view, reactant A is more likely to find its reaction partner B when the entire solution is filled with other B molecules, than when the solution contains virtually no B molecules. In other words, rates of reaction increase as the concentration of the reactants increase. But is the increase linear or exponential? That is what we wish to find out.

The specific reaction to be studied today involves the oxidation of iodide ion by hydrogen peroxide, which proceeds according to the following net stoichiometric reaction:

$$H_2O_{2(aq)} + 2\,I^-_{(aq)} + 2\,H^+_{(aq)} \longrightarrow I_{2(aq)} + 2\,H_2O$$

The reaction likely occurs in a series of steps as the probability of five species colliding at exactly the same time with sufficient energy and proper geometry is very small. One step is probably much slower than the others, and determines how fast the reaction proceeds.

The general rate law expression for the reaction of hydrogen peroxide with iodide ion in an acidic solution has the form:

$$\text{Rate} = -\,\Delta[H_2O_2]/\Delta t = k_{rate}\,[H_2O_2]^x\,[I^-]^y\,[H^+]^z$$

(13.1)

where $\Delta[H_2O_2]/\Delta t$ denotes the change in the concentration of hydrogen peroxide over time, and k_{rate} is the rate constant. The negative sign in front of $\Delta[H_2O_2]/\Delta t$ means that the hydrogen peroxide concentration decreases with time. The exponents "x", "y" and "z" (called reaction orders) are typically small whole numbers (usually, 0, 1 or 2). The objective of today's experiment will be to determine the numerical value of the rate constant, and the order of the chemical reaction with respect to hydrogen peroxide (numerical of x), iodide (numerical value of y) and hydrogen ion (numerical value of z).

A variation of the initial rate method will be used to measure the rate of reaction. A small amount of sodium thiosulfate ($Na_2S_2O_3$) and starch indicator is added to the solution. The thiosulfate ion does not readily react with any of the reactants; however, it does react very rapidly with the iodine

$$2\ S_2O_3{}^{2-}{}_{(aq)}\ +\ I_{2(aq)}\ ----\!\!\!-\!\!\!-\!\!\!>\ 2\ I^-{}_{(aq)}\ +\ S_4O_6{}^{2-}{}_{(aq)}$$

produced from the hydrogen peroxide reaction. The reaction with iodine is so fast that iodine is reduced back to the iodide ion as fast as it is formed. Once all of the thiosulfate is consumed the $I_{2(aq)}$ reacts with the starch indicator to form an intense blue complex. The time required to consume a fixed amount of $S_2O_3{}^{2-}$ is very reproducible.

How does this allow us to determine the rate law for the oxidation of hydrogen peroxide with iodide ion? Well, there is a mathematical relationship between how fast hydrogen peroxide and the thiosulfate ion are consumed. When the thiosulfate ion is present in solution, both chemical reactions are occurring simultaneously. By adding the two chemical reactions together, one finds that

$$H_2O_{2(aq)}\ +\ 2\ S_2O_3{}^{2-}{}_{(aq)}\ +\ 2\ H^+{}_{(aq)}\ ----\!\!\!-\!\!\!-\!\!\!>\ 2\ H_2O\ +\ S_4O_6{}^{2-}{}_{(aq)}$$

$$\text{Rate} = -\ \Delta[H_2O_2]/\Delta t\ =\ -\ 0.5\ \Delta[S_2O_3{}^{2-}]/\Delta t \tag{13.2}$$

The factor 0.5 appears in Eqn. 13.2 because hydrogen peroxide disappears half as fast as thiosulfate as shown by the coefficients in the two balanced equations.

The strategy for determining the rate law expression is quite simple. A series of solutions of known initial molar concentrations of $S_2O_3{}^{2-}$, I^- and H^+ will be prepared, and the time that it takes for the solution to turn intense blue will be noted. This will give a series of equations:

Solution 1: $-0.5\ \Delta[S_2O_3{}^{2-}]/\Delta t = k_{rate}[H_2O_2]^x_{solut1}[I^-]^y_{solut1}[H^+]^z_{solut1}$

$$\tag{13.3}$$

Solution 2: $-0.5\ \Delta[S_2O_3{}^{2-}]/\Delta t = k_{rate}[H_2O_2]^x_{solut2}[I^-]^y_{solut2}[H^+]^z_{solut2}$

$$\tag{13.4}$$

Solution 3: $-0.5\ \Delta[S_2O_3{}^{2-}]/\Delta t = k_{rate}[H_2O_2]^x_{solut3}[I^-]^y_{solut3}[H^+]^z_{solut3}$

$$\tag{13.5}$$

Solution 4: $-0.5 \, \Delta[S_2O_3^{2-}]/\Delta t = k_{rate}[H_2O_2]_{solut4}^{x}[I^-]_{solut4}^{y}[H^+]_{solut4}^{z}$

$$(13.6)$$

in which all of the numerical values will be known, except for the rate constant (k_{rate}) and the three reaction orders (x, y and z). The hydrogen ion concentration will be kept fixed by using a acetic acid + sodium hydroxide buffer system. The iodide ion concentration will be kept fixed because $I^-_{(aq)}$ is rapidly regenerated by the reaction of $I_{2(aq)}$ with $S_2O_3^{-2}{}_{(aq)}$ as fast as it is consumed. A nearly constant $[H_2O_2]$ is maintained by using a fairly large excess of hydrogen peroxide, relative to the amount of thiosulfate ion consumed.

In order to simplify the mathematical calculations, the solutions that will be studied will have two of the three reactant's concentrations the same, so that the respective concentrations will cancel when the correct ratio of two solution equations is used. Table 13.1 describes the preparation of the solutions to be studied, and Table 13.2 gives the initial concentrations of the reactants. For example, if one were to take the ratio of the equation for the first solution divided by the equation for the second solution, the resulting mathematical equation is:

$$\frac{\Delta \, time \, solution \, 2}{\Delta \, time \, solution \, 1} = (\frac{0.0533}{0.0267})^x = (2)^x \qquad (13.7)$$

The order of the chemical with respect to H_2O_2 is determined by the ratio of the time that it takes for Solution 2 to turn intense blue, divided by the time for Solution 1. Similarly, the reaction orders for I^-

$$\frac{\Delta \, time \, solution \, 3}{\Delta \, time \, solution \, 1} = (\frac{0.00833}{0.0167})^y = (0.5)^y \qquad (13.8)$$

and for H_3O^+

$$\frac{\Delta \, time \, solution \, 4}{\Delta \, time \, solution \, 1} = (\frac{1.75 \times 10^{-5}}{1.75 \times 10^{-4}})^z = (0.1)^z \qquad (13.9)$$

are determined in similar fashion. Once the three reaction orders are known, the numerical value of the rate constant, k_{rate}, is calculated from any of the four solution equations. One does need to know, however, the change in the thiosulfate ion concentration that occurred over the reaction time period. For all practical purposes, all of the thiosulfate was consumed, so the value of $\Delta[S_2O_3^{2-}]$ that would be substituted is $\Delta[S_2O_3^{2-}] = -0.00167$ Molar.

Table 13.1. Preparation of Reaction Mixtures to be Studied

Solution	Water	Buffer[a]	0.3 M CH_3COOH	0.05 M KI	1 % Starch	0.05 M $Na_2S_2O_3$	0.8 M H_2O_2[b]
1	75 ml	30 ml	0 ml	25 ml	5 ml	5 ml	10 ml
2	80 ml	30 ml	0 ml	25 ml	5 ml	5 ml	5 ml
3	50 ml	30 ml	0 ml	50 ml	5 ml	5 ml	10 ml
4	30 ml	30 ml	45 ml	25 ml	5 ml	5 ml	10 ml

[a]Buffer is 0.05 Molar acetic acid + 0.05 Molar sodium acetate.

[b]A 3 % hydrogen peroxide solution sold in drug stores is about 0.8 Molar.

Table 13.2. Initial Molar Concentrations of Chemical Species in the Reaction Mixtures

Solution	$[H_2O_2]$	$[I^-]$	$[H^+]$
1	0.0533	0.00833	1.75×10^{-5}
2	0.0267	0.00833	1.75×10^{-5}
3	0.0533	0.0167	1.75×10^{-5}
4	0.0533	0.00833	1.75×10^{-4}

REACTION TIME MEASUREMENTS

The Pasco system will be used as a timer in this particular experiment. Enter the Pasco system by **mouse clicking on the DataStudio icon** on the computer screen. When the screen saying "How would you like to use DataStudio" comes up, **mouse click on Create Experiment**. Plug the temperature probe into the PowerLink unit. Do not worry about calibrating the temperature probe. A probe or sensor has to be plugged into the unit in order to use to use the system as a timer.

The four reaction mixtures will be studied one at a time, starting with solution 1. Transfer by pipette or graduated cylinder all of the reactants except for hydrogen peroxide (H_2O_2) into a 200 or 250-ml beaker. Then using a pipet add exactly 10.0 ml of H_2O_2 and stir continuous with a glass stirring rod or plastic coffee stirrer (or a magnetic stirrer with stirring bar if available). When half of the H_2O_2 has drained out of the pipette, **mouse click on the Start button**. Do not worry about what is happening with the temperature column. You only need the time measurement. Continue stirring until you are sure that all of the solution is thoroughly mixed. Watch the solution carefully for the sudden appearance of the blue iodine-starch complex. Stop the timer when the solution turns blue. Record the time on the laboratory Data Sheet. The reaction time for the first solution should be between 2 and 4 minutes.

Measure the reaction times for solutions 2, 3 and 4 in the same manner. Be sure to add the hydrogen peroxide last. For each reaction, the timer should be started when half of the hydrogen peroxide has been added. The timer is stopped when the solution turns blue.

After all of the experimental data has been measured and recorded, calculate the rate constant and orders of the reaction with respect to H_2O_2, I^- and H^+ using Eqns. 13.3 – 13.9. You can safely assume that each reaction order is an integer value. The three possibilities for x, y and z are 0, 1 and 2. The measured reaction times do have experimental uncertainties associated with them. An error in one or more reaction volume will affect the times. Do not expect the calculated reaction time ratios to be perfect. For example, suppose that the reaction times for solutions 1 and 2 were 170 and 380 seconds, respectively. The calculated reaction time ratio (Eqn. 13.7) would be

$$\frac{\Delta\, time\, solution\, 1}{\Delta\, time\, soluiton\, 2} = (\frac{380\, sec}{170\, sec}) = 2.235 = (2)^x \qquad (13.10)$$

Of the values x = 0, x = 1 and x = 2, which one comes closest to giving 2^x = 2.235? The value 2^1 comes closest to 2.235. The reaction would then be first order in hydrogen peroxide. Analyze the experimental data in this manner. The value of the change in the thiosulfate concentration needed in the k_{rate} computation is $\Delta[S_2O_3^{2-}]$ = -0.00167 Molar.

DATA SHEET – EXPERIMENT 13

Name: _____

Reaction Time Measurement:

Solution 1

1. Time on timer when hydrogen peroxide is added, sec: _____

2. Time on timer when starch indicator turns blue, sec: _____

Solution 2

3. Time on timer when hydrogen peroxide is added, sec: _____

4. Time on timer when starch indicator turns blue, sec: _____

Solution 3

5. Time on timer when hydrogen peroxide is added, sec: _____

6. Time on timer when starch indicator turns blue, sec: _____

Solution 4

7. Time on timer when hydrogen peroxide is added, sec: _____

8. Time on timer when starch indicator turns blue, sec: _____

Rate Law Expression Results:

9. Reaction order with respect to hydrogen peroxide (value of x):   _____

10. Reaction order with respect to iodide ion (value of y):   _____

11. Reaction order with respect to hydrogen ion (value of z):   _____

12. Numerical value of the rate constant, k_{rate}:   _____

13. Units for the rate constant: _____

Mechanistic Interpretation:

Given below is a possible three-step mechanism describing the reaction of hydrogen peroxide with iodide ion in an acidic solution.

Step 1: (slow) $H_2O_{2(aq)} + I^-_{(aq)} \longrightarrow OH^-_{(aq)} + HOI_{(aq)}$

Step 2: (very fast) $OH^-_{(aq)} + H^+_{(aq)} \longrightarrow H_2O_{(liq)}$

Step 3: (fast) $HOI_{(aq)} + H^+_{(aq)} + I^-_{(aq)} \longrightarrow I_{2(aq)} + H_2O_{(liq)}$

Assuming that the above reaction mechanism is correct, write the expected rate law expression. Is your experimental data consistent with the proposed mechanism? Explain your answer.

EXPERIMENT 14: INTRODUCTION TO ACID-BASE CHEMISTRY – DETERMINATION OF ACETIC ACID CONCENTRATION IN VINEGAR

INTRODUCTION

Volumetric methods of analysis, also referred to as titrimetric methods, are widely used in analytical chemistry to determine the concentrations of unknown solutions. The method involves a chemical reaction between the substance to be quantified and a solution of a reagent of known concentration:

Substance + Reagent Solution -----> Products

The reagent solution of known concentration is called the titrant (the titrating agent) and the substance whose concentration we wish to determine is called the analyte. The titrant is added by buret until the analyte is completely consumed. The addition of titrant is immediately stopped at the neutralization point when all of the analyte is gone. At the neutralization point (equivalence point) there is neither an excess of titrant nor unreacted analyte. Both have been completely consumed. Using the stoichiometric coefficients in the balanced chemical reaction, the concentration and volume of titrant added, and the amount of unknown analyzed, the concentration of analyte in the unknown solution can be determined. Volumetric titrations do require that the titration reaction must be quantitatively complete, there must be a well-defined stoichiometric relationship between the analyte and titrant, and there must be some means to know when the equivalence point is reached. The latter is often accomplished using indicators, or dyes, whose color changes when the equivalence point is reached.

In Experiment 14 the concentration of acetic acid in a commercial sample of vinegar will be determined by an acid-base titration. Acetic acid (CH_3COOH) is a weak monoprotic acid that reacts with the strong base sodium hydroxide (NaOH):

$$CH_3COOH_{(aq)} + NaOH_{(aq)} \ \text{------>} \ H_2O + Na^+_{(aq)} + CH_3COO^-_{(aq)}$$

to form water plus the dissolved sodium acetate salt, $e.g.$, $Na^+_{(aq)}$ and $CH_3COO^-_{(aq)}$ ions. At the equivalence point:

moles of acetic acid = moles of sodium hydroxide   (14.1)

163

The number of moles of a chemical can be calculated as the mass of the chemical divided by the chemical's molar mass, or using the definition of molarity, one can express the number of moles as equal to the product of the molarity of the chemical times the volume of the chemical/solution (in liters). How we wish to calculate the number of moles in Eqn. 14.1 depends on what information is known and what quantity is to be calculated. In this instance, the molarity and volume of sodium hydroxide (the titrant) is known. The volume of acetic acid solution will also be known. The laboratory procedure calls for 5.00 ml of vinegar, which will be the volume of the acetic acid solution (vinegar) analyzed. Making the fore-mentioned substances into Eqn. 14.1, one obtains:

$$\text{Molarity of } CH_3COOH \times \text{Volume of } CH_3COOH = \text{Molarity of NaOH} \times \text{Volume NaOH}$$
(14.2)

Every quantity in Eqn. 14.2 will be known, except for the molarity of CH_3COOH.

The accurate quantification of the amount of acetic acid in vinegar depends on one stopping the addition of NaOH at the exact equivalence point. We need to find some way to determine the equivalence point. Indicator dyes provide a convenient means to signal the equivalence point in acid-base titrations. The indicator dye is usually an organic molecule whose color depends on whether or not the molecule is protonated. For example, in the protonated form, Hin, the dye might appear yellow in color, and in the unprotonated form, In⁻, the dye might appear red. There are many acid-base indicator dyes for possible use, and the key is to select the appropriate dye for the acetic acid titration. In chemistry reference books (*e.g.*, *CRC Handbook of Physics and Chemistry*, *Lange's Handbook of Chemistry*, etc.) one can find listing of possible acid-base indicator dyes and the pH range when the dye changes color. We know that in the titration of acetic acid with sodium hydroxide, that the solution will be acidic, pH < 7.0 prior to the neutralization point, and when there is an excess of sodium hydroxide added, the pH > 7.0. A pH = 7.0 corresponds to a neutral solution. Water does undergo self-dissociation

$$2 H_2O \quad \longleftrightarrow \quad H_3O^+_{(aq)} + OH^-_{(aq)}$$

to form the hydronium, H_3O^+, and hydroxide, OH^-, ions. The equilibrium constant for this equilibrium:

$$K_w = 1 \times 10^{-14} = [H_3O^+][OH^-]$$
(14.3)

has a numerical value of 1.0×10^{-14} at 25 °C. In deionized water, the only source of hydronium and hydroxide ions is from the dissociation equilibria. One hydroxide ion is produced for every hydronium ion, hence $[H_3O^+] = [OH^-]$. In a neutral solution of distilled water, one can substituted $[H_3O^+] = [OH^-]$ into Eqn. 14.3, to obtain $[H_3O^+] = [OH^-] = 1 \times 10^{-7}$. Since pH is defined to be pH $= -\log [H_3O^+]$, a neutral solution would have pH $= -\log 10^{-7} = 7.0$. Our solution will not be neutral at the equivalence point,

164

because the titration reaction produces the acetate ion, which is the conjugate base of acetic acid. Even in the absence of excess NaOH the solution would be slightly basic at the equivalence point due to the acetate ions that are produced. In a later laboratory experiment we will learn how to calculate the pHs for various solutions; however, for now, we will simply assume that an indicator having a color change at a pH slightly above 7 will suffice. Of the indicators listed in Table 14.1, phenolphthalein is the best choice amongst the common pH indicators. Phenolphthalein will change from colorless to faint pink at the equivalence point (titration endpoint).

Examination of Table 14.1 shows that the indicators do not change at a specific pH value, but rather gradually over a pH range. The observed color depends on the ratio of the protonated form (acid form) to unprotonated form (base form) of the indicator. Generally a concentration ratio of somewhere between 8:1 to 10:1 is needed before the color is distinct. Take for example bromocresol green. One would need the concentration of the protonated form of the indicator to be 8 to 10 times greater than the concentration of the unprotonated form in order to see a distinct yellow color. Similarly, a distinct blue color would be observed if the unprotonated form was 8 to 10 times that of the protonated form. The solution would likely appear green whenever the two indicator form concentrations were about equal.

Table 14.1. Transition Ranges and Colors of Common Acid-Base Indicators[†]

Indicator	pH range	Acid	Base
Thymol blue*	1.2 – 2.8	Red	Yellow
Methyl yellow	2.9 – 4.0	Red	Yellow
Methyl orange	3.1 – 4.4	Red	Yellow
Bromocresol green	3.8 – 5.4	Yellow	Blue
Methyl red	4.2 – 6.2	Red	Yellow
Bromocresol purple	5.2 – 6.8	Yellow	Purple
Thymol blue*	8.0 – 9.6	Yellow	Blue
Phenolphthalein	8.0 – 9.8	Colorless	Pink
Alizarin yellow	10.0 – 12.1	Yellow	Orange-red

*Has two pH transition ranges.
† A more complete listing of acid-base indicators can be found in standard analytical chemistry books (for example, see Handbook of Analytical Chemistry, 1st edition, L. Meites, McGraw-Hill, New York, NY (1963); *Fundamentals of Analytical Chemistry*, 7th edition, by D. A. Skoog, D. M. West and F. J. Holler, Saunders College Publishing, New York, NY (1996); *Quantitative Chemical Analysis*, 6th Edition, by Daniel C. Harris, W. H. Freeman and Company (2003)).

One additional note about indicators. In this experiment the equivalence point is located with a commercial chemical indicator. Alternative, a "nature indicator" could have been used. For example, red cabbage, grapes, blackberries, cherries, cranberries and plums contain pigment molecules called anthocyanins. Very acidic solutions will turn anthocyanins a red color, neutral solutions have a purplish color, and basic solutions will appear greenish-yellow. The colors and approximate transitions for red cabbage are: deep red at pH = 1 to purple at pH = 7 to green at pH = 12, and at about pH = 13.5 the pigments turn yellow. Cherries and cherry juice is bright red in acid solutions, but purple to blue in basic solutions. Olfactory indicators (onion and garlic) are also available. For these indicators the endpoint would be detected by changes in smell, rather than changes in color. Onion and garlic odor are both quite noticeable in acidic solutions, and virtually undetectable in strongly basic solutions. A more detailed discussion of olfactory indicators is found in K. Neppel, M. T. Oliver-Hoyo, C. Queen and N. Reed, J. *Chem. Educ.*, **82**, 607-610 (2005).

TITRATION OF VINEGAR USING PHENOLPHTHALEIN AS INDICATOR

Add exactly 5.00 ml of vinegar with a pipet or buret to a 400-ml (or similar size) Erlenmeyer flask. Record the volume of vinegar used on the laboratory Data Sheet under the column heading labeled "Trial 1". Add about 50 ml of deionized water (graduated cylinder) to the Erlenmeyer flask. Next add two or three drops of phenolphthalein indicator to the Erlenmeyer flask. The solution is acidic and the indicator should be in its protonated form (colorless).

Next you need to rinse the burette with about 10 ml of the sodium hydroxide titrant that you are going to use. This can be accomplished by lowering the burette so that the top is easy to reach and carefully pouring the titrant into the top (be sure the stopcock is in the closed position), or if you prefer add the titrant with a small funnel. It is easier to fill the burette using a funnel, but as you will learn in analytical chemistry, a funnel increases the likelihood that the titrant may become contaminated. For the introductory level course, contamination is not so much a concern as you are just learning the experimental technique, and the grade on the laboratory experiment is not based very heavily on how close your measured value is to the so-called true or accepted value. Using a small funnel is fine. Drain the titrant into a waste container.

Fill the burette with 0.100 Molar (or similar molarity) sodium hydroxide to an appropriate level so that a volume reading can be made. It is good practice of "overfill" the burette slightly, and then allow the liquid to run into a waste container until the desired start reading is reached. **ALWAYS INSPECT THE TIP OF THE BURETTE TO MAKE SURE THAT THERE IS NO "AIR POCKET" THAT COULD BECOME DISLODGED DURING THE COURSE OF THE TITRATION.** If you do notice an air pocket, open the stopcock slightly and allow a few ml of titrant to drain. This will generally eliminate the air pocket. If not, gently tap on the tip of the burette while the titrant is draining. This often helps. Before you begin the titration the **air pocket must be removed**. Otherwise, the titrant will displace the air procket during the titration, and not all of the titrant used will be delivered to the titration flask. Some will remain in the space that was once occupied by the air pocket. Record the molarity of sodium hydroxide and the initial buret reading on the Data Sheet. For clear liquids, the burette reading is taken as the bottom of the U-shaped meniscus. Now you need to start the titration. (See Figure 14.1 for set up of apparatus.) When you get near the endpoint you will need to be adding the titrant (NaOH) dropwise. As far as you know it may take a large volume of NaOH to reach the endpoint, and adding the titrant dropwise will take a lot of time. One can tell when the endpoint is near by carefully observing the solution near where the NaOH is dripping. When you get close to the endpoint you will notice that the solution turns pink around where the NaOH solution falls. The closer you get to the endpoint, the longer the pink color persists. You need to adjust the stopcock on the buret to dropwise addition when you start to observe the solution turning pink where the NaOH solution falls. The initial pink color will disappear quickly as the solution is swirled, and the NaOH reacts with the unreacted acetic acid that remains in the rest of the solution. The Erlenmeyer flask needs to be gently swirled as the sodium hydroxide is added in order to completely mix the contents. Be sure to titrate slowly and carefully as the endpoint is approached. The endpoint occurs when the entire solution turns faint pink, and the pink color persists for at least 30 seconds. Record the final buret reading on the Data Sheet. Now repeat the entire experiment over again, and entire the experimental data on the Data Sheet under the column heading labeled "Trial 2".

Do not be surprised if the solution turns back colorless upon sitting for several minutes. At the endpoint the solution is slightly basic. If the solution is left open to the atmosphere, it will absorb carbon dioxide from the air. Carbon dioxide dissolved in water forms carbonic acid, which will then react with the hydroxide ions to form water. Once sufficient carbon dioxide has dissolved in the solution, the solution pH is below the transition point for the indicator to appear pink, and the solution will then go back to colorless. [Note: The dianion of phenolphalein (pink form) does react with the hydroxide ion at higher pHs (see J. R. Lalanne, *J. Chem. Educ.*, **48**, 266-268 (1971)). In our case the fading of the pink color is due to carbon dioxide adsorption as the pH of the titration solution should be considerably less than pH = 11, assuming that one stopped the titration at the very first sign of permanent pink color.]

Calculate the molarity of acetic acid in vinegar by substituting the molarity and volume of NaOH, and the volume of vinegar analyzed into Eqn. 14.2. How does this value compare to the amount of acetic acid that is labeled on the bottle? To do this comparison, you need to convert the molarity that you determined into a mass/volume percentage. Mass/volume percentage is defined as 100 times the mass of the solute (in grams) divided by the volume of the solution analyzed (in ml). The "100" converts the value to percent. Molarity is converted to mass/volume percent by

$$\text{mass/volume \%} = 100 \times \text{molarity of CH}_3\text{COOH} \times (\text{molar mass of CH}_3\text{COOH}/1000)$$

(14.4)

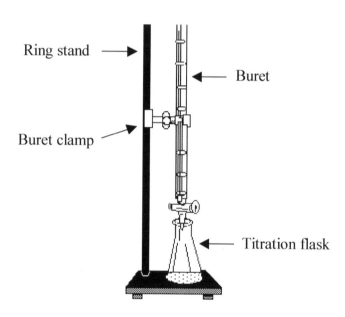

Figure 14.1. Titration apparatus for the determination of acetic acid in vinegar.

TITRATION OF VINEGAR USING METHYL ORANGE AS INDICATOR

It has already been pointed out that the pH at the equivalence point will be slightly greater than 7. At the equivalence point the solution will be basic, and one needs to select an appropriate indicator having a color change in this pH region. But let's

168

suppose that phenolphthalein was not available, and that the only indicator that we had on hand was methyl orange, which changes color in the pH 3.1-4.4 range. Using methyl orange has the indicator would be cause a systematic error in the analysis (*e.g.*, error in experimental procedure, as opposed to a random indeterminant error). Would the error be significant? Let's find out. Perform the acetic acid – NaOH titration again, this time substituting three drops of methyl orange for phenolphthalein. Record the experimental data on the Data Sheet in the section titled "Acetic Acid Titration with Methyl Orange Measurements." As before calculate the molarity and mass/volume percentage of acetic acid in vinegar.

TITRATION OF VINEGAR USING BROMOCRESOL GREEN AS INDICATOR

In the preceding titration you likely observed that it took quite a bit less sodium hydroxide when methyl orange was used. Perhaps one could find a slightly better indicator than methyl orange, an indicator that changes colors a little bit closer to a pH slightly greater than 7. Bromocresol green changes color in the pH 3.8 – 5.4 range. Perform the acetic acid titration one last time, this time substituting three drops of bromocresol green for phenolphthalein. Record the experimental data on the Data Sheet in the section titled "Acetic Acid Titration with Bromocresol Green Measurements." As before calculate the molarity and mass/volume percentage of acetic acid in vinegar. Give your mass/volume percentage experimental values to the TA who will then compile and post the entire class's data for the phenolphthalein, methyl orange and bromocresol green titrations.

CLEANING PROCEDURE FOR BURETTE

At the end of the experiment it is important to rinse the burette several times with distilled water to remove any titrant from the burette. Remove the stopcocks and rinse the stopcock assembly thoroughly. Reassemble the burette and rinse with deionized water. Invert the burette in the burette holder (that is, hang the burette upside down) with the stopcock open to dry. This will prevent the tip from getting clogged with a solid. If the burette is not thoroughly cleaned the tip will likely become clogged with solid once the water has evaporated. Leave the stopcock in this configuration for the next laboratory session.

COMPARISON OF PHENOLPHTHALEIN, METHYL ORANGE AND BROMCRESOL GREEN TITRATION RESULTS

Based solely upon your experimental measurements, there is really not a sufficient number of data points to do much in the way of a statistical comparison of the two different analytical methods, phenolphthalein indicator versus methyl orange indicator, and phenolphthalein versus bromocresol green indicator. By pooling the entire class's experimental data, the number of experimental observations can be significant increased. Pooling the data does introduce a few additional variables, such as the color perception of different individuals, *etc.* For illustrating the statistical method this will have to suffice as there is not sufficient time in the laboratory for you to make the necessary number of replicate measurements to do a Student t-test statistical analysis. The TA will gather the experimental mass/volume percentages for the phenolphthalein, methyl orange and bromocresol green indicator titrations, and will post the values on the bulletin board outside the laboratory room sometime during the next day. Go to the bulletin board outside the laboratory room and copy down the class results on your laboratory Data Sheet in the space provided. Calculate the class mean (average) and standard deviation for the three data sets.

The Student's t-test will be used to decide whether there is a significant difference between any two methods. The Student's t-value is calculated according to Eqn. 14.5

$$t_{calc} = \frac{|\bar{x}_1 - \bar{x}_2|}{s_{pooled}} \sqrt{\frac{n_1 \, n_2}{n_1 + n_2}} \tag{14.5}$$

where

$\bar{x}_1$ = mean value for the phenolphthalein titration

$\bar{x}_2$ = mean value for the methyl orange (or bromocresol green) titration

n_1 = number of measured values for the phenolphthalein titration

n_2 = number of measured values for the methyl orange (or bromocresol green) titration

s_{pooled} = pooled standard deviation.

The value of s_{pooled} is calculated using the standard deviations of the two sets of titration, s_1 and s_2, respectively, according to Eqn. 14.6:

$$s_{pooled} = \sqrt{\frac{s_1^2 (n_1 - 1) + s_2^2 (n_2 - 1)}{n_1 + n_2 - 2}} \tag{14.6}$$

You may wish to refer back to Experiment 1 for a more detailed discussion of the Student's t-test. Table 14.2 lists the theoretical values of Student's t at the 95 % confidence level for several different values of $n_1 + n_2$. If your number of measurements is not listed in the table, use the closest value. If the value of t_{calc} that you calculate based on your experimental data exceeds the value given in Table 14.2, then there is a difference in the two data sets considered for the 95 % level of confidence that you were testing at. If t_{calc} does not exceed the tabulated value in Table 14.2, then there is not a significant difference in the two data sets.

The values in Table 14.2 are listed according to the total number of measurements performed. In many texts, the values would be tabulated according to the "degrees of freedom". For a comparison of experimentally determined average values for two data sets, as is the case here, the number of degrees of freedom is $n_1 + n_2 - 2$. The t-test also can be used to compare an experimental average value to a so-called "known" value, as one might want to do in determining whether or not an instrument was properly calibrated. In the later statistical analysis (average value compared to a "known" value) the number of degrees of freedom would be $n - 1$. There would be only the single data set.

Table 14.2. Student's t for comparing two data sets at the 95 % confidence level

Measurements	Student's t	Measurements	Student's t
4	4.303	19	2.110
5	3.182	20	2.101
6	2.776	21	2.093
7	2.571	22	2.086
8	2.447	23	2.080
9	2.365	24	2.074
10	2.306	25	2.069
11	2.262	26	2.064
12	2.228	27	2.060
13	2.201	28	2.056
14	2.179	29	2.052
15	2.160	30	2.048
16	2.145	31	2.045
17	2.131	32	2.042
18	2.120		

A more complete listing of Student t-values can be found in standard analytical chemistry textbooks (see for example, *Fundamentals of Analytical Chemistry*, 7th edition, by D. A. Skoog, D. M. West and F. J. Holler, Saunders College Publishing, New York, NY (1996); *Quantitative Chemical Analysis*, 6th Edition, by Daniel C. Harris, W. H. Freeman and Company (2003)), standard statistics books (for example, *Introduction to the Theory of Statistics*, 3rd edition, A. M. Mood, F. A. Grayhill and D. C. Boes, McGraw-Hill, Inc., New York, NY (1974); *Methods of Statistical Analysis*, 2nd edition, C. H. Goulden, Wiley, New York, NY (1956)) and in science and engineering handbooks (see for example *Lange's Handbook of Chemistry*, 13th edition, by J. A. Dean, McGraw-Hill, New York, NY (1985); CRC Standard Mathematical Tables, 28th edition, W. H. Beyer, CRC Press, Boca Raton, FL (1987)).

DATA SHEET – EXPERIMENT 14

Name: _____

acetic acid

Acetic Acid Titration with Phenolphthalein Measurements:

	Trial 1	Trial 2
1. Volume of vinegar analyzed, ml:	_____	_____
2. Molarity of NaOH used, M:	_____	_____
3. Initial buret reading for NaOH, ml:	_____	_____
4. Final buret reading for NaOH, ml:	_____	_____
5. Volume of NaOH used, ml:	_____	_____
6. Volume of NaOH used, l:	_____	_____
7. Molarity of acetic acid in vinegar, M:	_____	_____
8. % Mass/Volume of acetic acid in vinegar:	_____	_____
9. Average % Mass/Volume of acetic acid:	_____	
10. % Mass/Volume of acetic acid (on label):	_____	

Acetic Acid Titration with Methyl Orange Measurements:

11. Volume of vinegar analyzed, ml: _____

12. Molarity of NaOH used, M: _____

13. Initial buret reading for NaOH, ml: _____

14. Final buret reading for NaOH, ml: _____

15. Volume of NaOH used, ml: _____

16. Volume of NaOH used, l: _____

17. Molarity of acetic acid in vinegar, M: _____

18. % Mass/Volume of acetic acid in vinegar: _____

19. % Mass/Volume of acetic acid (on label): _____

Acetic Acid Titration with Bromocresol Green Measurements:

20. Volume of vinegar analyzed, ml: _____

21. Molarity of NaOH used, M: _____

22. Initial buret reading for NaOH, ml: _____

23. Final buret reading for NaOH, ml: _____

24. Volume of NaOH used, ml: _____

25. Volume of NaOH used, l: _____

26. Molarity of acetic acid in vinegar, M: _____

27. % Mass/Volume of acetic acid in vinegar: _____

28. % Mass/Volume of acetic acid (on label): _____

Class Data – Mass/volume percentage from phenolphthalein titration:

_____	_____	_____	_____
_____	_____	_____	_____
_____	_____	_____	_____
_____	_____	_____	_____
_____	_____	_____	_____
_____	_____	_____	_____

Class Mean for phenolphthalein titration: _____

Class Standard deviation for phenolphthalein titration: _____

Class Data – Mass/volume percentage from methyl orange titration:

_____	_____	_____	_____
_____	_____	_____	_____
_____	_____	_____	_____
_____	_____	_____	_____
_____	_____	_____	_____
_____	_____	_____	_____

Name: _____

Class Mean for methyl orange titration: _____

Class Standard deviation for methyl orange titration: _____

Class Data – Mass/volume percentage from bromocresol green titration:

_____	_____	_____	_____
_____	_____	_____	_____
_____	_____	_____	_____
_____	_____	_____	_____
_____	_____	_____	_____
_____	_____	_____	_____

Class Mean for bromocresol green titration: _____

Class Standard deviation for bromocresol green titration: _____

Comparison of the two analytical methods (comparison of phenolphalein versus methyl orange indicators):

What is the numerical value of s_{pooled}? _____

What is the numerical value of t_{calc}? _____

Statistically, two data sets are "considered different" if they are different at the 95 % confidence level or greater. Does your experimental data fulfill this criteria? In other words, does your value of t_{calc} exceed the value Table 14.2 for the total number of measurements made? _____

Comparison of the two analytical methods (comparison of phenolphalein versus bromocresol green indicators):

What is the numerical value of s_{pooled}? _____

What is the numerical value of t_{calc}? _____

Statistically, two data sets are "considered different" if they are different at the 95 % confidence level or greater. Does your experimental data fulfill this criteria? In other words, does your value of t_{calc} exceed the value Table 14.2 for the total number of measurements made? _____

EXPERIMENT 15: INTRODUCTION TO pH – TITRATION OF ACETIC ACID IN VINEGAR AND PHOSPHORIC ACID IN COCA-COLA

INTRODUCTION

In today's laboratory experiment you will study the dissociation behavior of the weak monoprotic acetic acid, CH_3COOH, and the weak triprotic phosphoric acid, H_3PO_4, by sodium hydroxide titration. The titration will be followed using a pH glass membrane electrode. The glass membrane pH electrode is a type of specific-ion electrode, in that the electrode responds specifically to H_3O^+ ions. The electrode's selectively is due to the special composition of the thin glass membrane that separates an internal reference H_3O^+ solution (inside the electrode) from the solution whose pH one wishes to measure. The membrane is specific in its response toward hydrogen ions up to a pH of about 9. At higher pH values, however, the glass membrane becomes somewhat responsive to sodium ion, as well as other singly charged cations. A detailed understanding of how the electrode functions, and why the membrane is specific H_3O^+ and not other ions, is beyond the scope of today's laboratory experiment, and would require a fundamental understanding of electrochemistry that is more in depth than what is typically covered in introductory general chemistry. All that you need to know for today's experiment is that the electrode measures the pH of the solution, and this allows you to calculate the H_3O^+ ion concentration

$$pH = - \log [H_3O^+] \tag{15.1}$$
$$[H_3O^+] = 10^{-pH} \tag{15.2}$$

for each incremental addition of sodium hydroxide. This information will be needed for you to calculate the numerical values of the acid dissociation constant(s) that describe the dissociation behavior of acetic acid and of phosphoric acid in water. Visual indicators like phenolphthalein and methyl orange, which were used in Experiment 14, can signal the equivalence point in an acid-base titration by color changes; however visual indicators do not give the hydronium ion concentration that is needed in acid dissociation constant calculations.

DISCUSSION OF TITRATION CURVE FOR A WEAK MONOPROTIC ACID – ACETIC ACID

Weak monoprotic acids, like acetic acid (CH_3COOH), only partially dissociate in water:

$$CH_3COOH_{(aq)} + H_2O \longleftrightarrow H_3O^+_{(aq)} + CH_3COO^-_{(aq)}$$

The bi-directional arrow, "$\longleftrightarrow$", indicates a dynamic equilibrium. As acetic acid molecules are transferring a proton to water to form the hydronium and acetate ions, other hydronium and acetate ions are colliding elsewhere in the solution to reform the acetic acid molecule. At equilibrium the rates of reaction in the forward and in the reverse directions are equal.

From a mathematical standpoint, the equilibrium condition for the dissociation of acetic acid is described by the equilibrium constant

$$K_a = \frac{[H_3O^+][CH_3COO^-]}{[CH_3COOH]} \tag{15.3}$$

which is a ratio of the product of molar concentrations of the products divided by the product of the molar concentrations of the reactants. By thermodynamic convention, the "pure" solids and the solvent do not appear in the equilibrium constant. The equilibrium constant is given the special subscript "a" (meaning "acid"), and called the acid dissociation constant.

The objective of today's laboratory experiment is to determine the actual numerical value of the equilibrium constant. Looks fairly simple, all that one has to do is substitute numerical values for the three quantities on the right-hand side of Eqn. 15.3. The determination of $[H_3O^+]$ is very straightforward. This value is calculated from the measured solution pH through Eqn. 15.2. The other two concentrations are not so easy to calculate in an absolute sense. Although the total amount of acetic acid that is in vinegar is known from last week's experiment, we do not know how much acetic acid has actually dissociated. In order to calculate the amount of acetic acid that actually dissociated one needs to know the acid dissociation constant. In order to calculate the acid dissociation one needs to know the fraction of acetic acid that dissociated. Initially it may appear that there is no way around this dilemma, fortunately there is, otherwise you would not be doing today's experiment.

Let us ask the following question, is there any set of special mathematical circumstances under which the ratio $[CH_3COO^-]/[CH_3COOH]$ can be canceled from Eqn. 15.3? Yes, if $[CH_3COO^-] = [CH_3COOH]$. Where would this point occur during the titration? When both concentrations are 50 % of the initial acetic acid concentration, *e.g.*, whenever the acetic acid is 50 % neutralized. The problem now becomes one of measuring the pH of the titration solution whenever the acetic acid has been 50 %

neutralized. In the titration of acetic acid (beaker) with sodium hydroxide (buret), the solution is initial acidic. As sodium hydroxide is added, the concentration of acetic acid decreases and the solution becomes less acidic. At the very beginning of the titration the decrease in acetic acid is not very noticeable as there is quite a bit of acetic acid in vinegar, and each ml of titrant contains a fixed, fairly small number of moles of sodium hydroxide. As more and more of the acetic acid has reacted, a one ml addition of sodium hydroxide still makes only a slight difference in the measured pH. For most of this time, one is in the "buffered" region of the titration curve. The largest difference occurs near the equivalence point, when two or three drops of titrant result in a sharp increase in the solution's pH. Once an excess of titrant is present, the incremental addition of still more titrant does not significantly change the solution's pH. A graph of the pH of the solution versus the volume of titrant added gives an S-shaped curve. The essentially flat portion of the curve is the "buffer" region of the titration curve. The equivalence point corresponds to the inflection point in the sharply increasing pH portion of the titration curve. A typical pH titration curve for the titration of acetic acid with sodium hydroxide is given in Figure 15.1. The "half-neutralization" point is found by first locating the equivalence point of the titration (the middle of the sharply increasing portion of the pH titration curve), and then dividing the volume of sodium hydroxide by 2. Next find the volume of sodium hydroxide at the half-neutralization point on the titration curve, and then go up to the curve, and horizontally straight across to get the measured pH value at the half-neutralization point on the y-axis. Convert the pH to $[H_3O^+]$ using Eqn. 15.2. Careful examination of Eqn. 15.3 reveals that $K_a = [H_3O^+]$ at the "half-neutralization" point.

Acid dissociation constants of weak acids are sometimes quite small, and out of mathematical convenience the numerical values are often reported in terms of their negative logarithmic value, e.g., $pK_a = -\log K_a$ (analogous to the definition of pH). Halfway to the equivalence point in a weak acid titration the measured pH = pK_a. Knowledge of the dissociation constants of weak acids and bases is important in the preparation of buffer solutions. Buffers usually consist of approximately equal quantities of a weak acid and its conjugate base, or a weak base and its conjugate acid. If one wished to prepare a buffered solution of lets say pH = 4, then one would use a weak acid whose pK_a is close to 4, and then the weak acid/conjugate base concentration would be adjusted until the desired pH is obtained. For doing quick calculations one can use the Henderson-Hasselbalch buffer equation:

$$pH = pK_a - \log\frac{[acid]}{[base]} \qquad (15.4)$$

which is obtained by taking the negative logarithm of both sides of Eqn. 15.3, and then isolating $- \log [H_3O^+]$ on one side of the equation.

The equivalence point of an acid-base titration is in the middle of the steeply rising pH portion of the titration curve, as depicted in Figure 15.1; *e.g.* the equivalence point occurs at the inflection point of the steeply rising region. If one were to place a pencil flat on the titration curve, and then move the pencil slowly from left to right along the titration curve, what do you notice about the slope that is defined by your pencil? In the flat, buffered region of the titration curve, the pencil is essentially horizontal with a slope of zero. As the pencil starts moving along the sharply increasing pH portion of the titration curve, the slope continues to increase for a while until the pencil is close to horizontal, and then gradually decreases back to zero as the moves to the part of the titration curve where the sodium hydroxide is in large excess. The inflection point occurs when the slope is at its maximum value. The equivalence point can be located by taking the derivative of the pH titration curve. The Pasco software has the capability of taking both the first and second derivative of a curve for you using its built-in software. While it may be easy to find the equivalence point (inflection point) in the acetic acid titration curve, the equivalence point is much harder to find in titration curves of polyprotic acids.

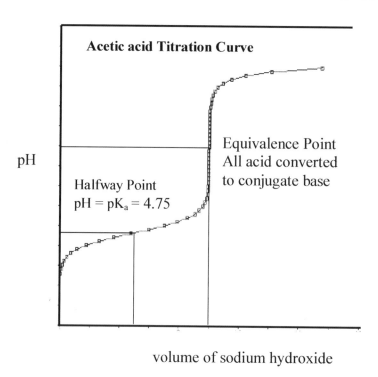

Figure 15.1. Titration curve of acetic acid with the strong base sodium hydroxide.

180

DISCUSSION OF TITRATION CURVE FOR A WEAK TRIPROTIC ACID – PHOSPHORIC ACID

The titration curve for the triprotic phosphoric acid is a bit more complicated in that the protons are transferred to water in stepwise fashion:

First proton transfer: $H_3PO_{4(aq)} + H_2O \longleftrightarrow H_3O^+_{(aq)} + H_2PO_4^-_{(aq)}$

$$K_{a1} = \frac{[H_3O^+][H_2PO_4^-]}{[H_3PO_4]} \qquad (15.5)$$

Second proton transfer: $H_2PO_4^-_{(aq)} + H_2O \longleftrightarrow H_3O^+_{(aq)} + HPO_4^{2-}_{(aq)}$

$$K_{a2} = \frac{[H_3O^+][HPO_4^{2-}]}{[H_2PO_4^-]} \qquad (15.6)$$

Third proton transfer: $HPO_4^{2-}_{(aq)} + H_2O \longleftrightarrow H_3O^+_{(aq)} + PO_4^{3-}_{(aq)}$

$$K_{a3} = \frac{[H_3O^+][PO_4^{3-}]}{[HPO_4^{2-}]} \qquad (15.7)$$

The subscript notation for the acid dissociation constant has been expanded to indicate whether the first proton (subscripted as "a1"), or the second proton (subscripted as "a2") or third proton (subscripted as "a3") has been transferred to water. From a mathematical standpoint, the stepwise dissociation equilibrium of a polyprotic acid can be treated independently if the values of the dissociation constants differ by a factor of 10^3 or more, as is the case with phosphoric acid. This means that sodium hydroxide essentially removes the first proton from all of the phosphoric acid before any of the second proton is removed. Similarly, the second proton is completely removed before the third proton is lost. The resulting titration curve then looks the combination of a series of three separate monoprotic titration curves, see Figure 15.2.

At the first, second and third equivalence points the predominant phosphoric acid species in solution is $H_2PO_4^-$, HPO_4^{2-}, and PO_4^{3-}, respectively. Halfway between any two equivalence points, the ratio of the molar concentration of the acid to its corresponding conjugate base is unity. Halfway to the first equivalence $[H_3O^+] = K_{a1}$; halfway between the first and second equivalence points $[H_3O^+] = K_{a2}$; and halfway between the second and third equivalence points $[H_3O^+] = K_{a3}$. Note that the volume of tritrant needed to remove the first proton equals the volume needed to remove the second, which is also equal to the volume needed to remove the third proton. In the pH region where the dissociation constants are determined, the pH curve is fairly flat. A two or three ml error in locating the halfway point should have only a very small effect on the numerical value of K_a. It is very unlikely that you will actually observe the third equivalence point in today's phosphoric acid titration. The molar concentration of sodium hydroxide is much too low to reach the pH needed to remove the third proton from all of the phosphoric acid initially present. You should be able to obtain a very reasonable value for the third dissociation constant, though, as the highest pH that will be reached is near the third horizontal region where K_{a3} is determined.

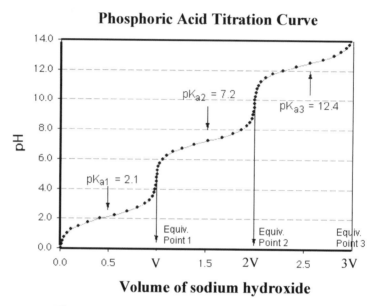

Figure 15.2. Titration curve for the titration of phosphoric with sodium hydroxide.

REMOVAL OF CARBON DIOXIDE FROM COCA COLA

Before beginning the phosphoric acid titration, you will first need to boil the Coca Cola to remove the dissolved carbon dioxide. Otherwise, you will be titrating both phosphoric acid and carbonic acid (formed from the dissolution carbon dioxide in aqueous solution). You will likely want to do this step first as it will take a while for the Coca Cola to cool back to room temperature. While the Coca Cola is cooling, you can perform other parts of the experiment, and by the time the Coca Cola is actually needed, it should have cooled substantially. Place 50 ml of Coca Cola in a beaker, cover the beaker with a watch glass, and boil gently (simmering will be fine) for about 20 minutes or until the volume of liquid is reduced by one-third, whichever comes first. Do not boil to dryness. After the Coca Cola has boiled, remove the source of heat, and let the solution cool, while you calibrate the Pasco system and do the acetic acid in vinegar titration.

182

SET UP AND CALIBRATION OF THE PASCO SYSTEM FOR A pH TITRATION MEASUREMENT

Enter the Pasco system by **mouse clicking on the DataStudio icon** on the computer screen that comes up when the computer is first turned on. When the screen saying "How would you like to use DataStudio" comes up, **mouse click on Create Experiment**. Connect the Pas*Port* pH probe and Pas*Port* Drop Counter accessory to the PowerLink unit. The system should recognize what accessories have been attached. Go to the top tool bar and **mouse click on the Setup button**. The setup window will now open up on the screen. The set up window will be for calibrating either the pH probe or the drop counter. One can go back and forth between the two calibrations by clicking on the ▼ or ▶ symbol to the left of the accessory name. On the screen for the pH/ISE/Temperature sensor, there will be three options displayed on the screen. You will want to **click on the Calibrate button directly across from the pH box**. The pH calibration is a two-point calibration. For the pH ranges that you will be measuring, the electrode should be calibrated using both an acidic pH standard (pH = 4 would be fine) and a basic pH standard (pH = 10 is fine). Place the pH electrode in the first pH standard/buffered solution, **type in the pH of the buffer in the box for Point 1**, wait a few minutes for the pH electrode to reach equilibrium with the solution, and then **mouse click on the Set button**. The first pH point is now set. Take the pH electrode out of the standard solution, hold the electrode over an empty beaker, rinse the electrode with deionized water from your wash bottle, and carefully blot dry with a paper towel, before placing in the second pH standard/buffer solution. Although the pH of a buffered solution is not suppose to change upon addition of a small amount of acid or base, it is good practice to rinse the pH electrode with deionized before placing it into the next solution. The pH of Point 2 is set by placing the pH electrode into the second standard/buffered solution, **typing in the pH in the box for Point 2**, waiting a few minutes for the electrode to reach equilibrium with the buffered solution, and then **mouse clicking on the Set button. Click the OK button** when both points have been set.

Now you need to set up and calibrate the Pasco Drop Counter. Attach the two plastic stopcocks one above the other onto the plastic syringe buret. The top stopcock will be used to adjust the drip rate. The bottom stopcock will be used to turn the titrant flow on (full vertical position) or off (full horizontal position). Once the drip rate is set **DO NOT TURN THE TOP STOPCOCK AGAIN.** This would change the drip rate. Fill the syringe with the deionized water. Open the bottom stopcock to the full vertical position. Now slowly turn the top stopcock until you get a steady drop rate of 3 to 4 drops per second. Once this is achieved, turn the bottom stopcock to the horizontal position. **DO NOT TURN THE TOP STOPCOCK AGAIN UNTIL YOU ARE READY TO ADJUST THE DRIP RATE.**

Now you need to tell the system how large the falling drops are. Position the syringe buret over the rectangular opening in the Drop Counter. Leave room under the drop counter to place the titration beaker and a graduated cylinder. Place a graduated cylinder under the Drop Counter, aligned with the buret so that the dripping titrant will fall into the graduated cyclinder. The cylinder will be used to measure the volume of the titrant that is delivered from the buret. **Mouse click on the Setup button** on the top toolbar. The setup window will now reappear on the screen. **Mouse click on the ▼ or ▶ symbol** to the left Drop Counter. The calibrate screen should now appear on the screen. **Mouse click on the Calibrate button.** Open the bottom stopcock to allow the liquid drops to fall directly into the graduated cylinder. You do not want the liquid to drops to hit the Drop Counter or to miss the beaker. If for some reason you need to turn off the buret before the calibration is finished, or if the drops miss the graduated cylinder, turn the bottom stopcock to the full horizontal off position. You will need to empty any liquid from the graduated cylinder, close the calibration screen, and then re-enter the Drop Counter calibration screen by mouse clicking on the Setup button on the top tool bar. Once the liquid in the graduated cylinder reaches the 10-ml mark, turn the bottom stopcock to the horizontal off position. You now need to tell the system how much liquid actually flowed between the Drop Counter cells. Since 10-ml of liquid was collected in the graduated cylinder, **type 10.00 in the box on the calibration screen. Mouse click on the Set button on the calibration screen.** The number that is above the box should change to 10.00 (*e.g.*, the number that you typed). When this has been done, **mouse click OK**. The Drop Counter should now be calibrated. The green light on the Drop Counter should be flashing the entire time the drops are falling. If the green light stays on (does not flash), this indicates that some liquid has splashed onto the lens. Should this happen gently position a paper towel into the rectangular box, and move along the inside walls to remove the splashed water.

After the pH electrode and Drop Counter have been calibrated, you will need to set the system up to display the measured data as a pH titration curve, *i.e.*, pH (y-axis) versus volume of titrant (x-axis) plot. Close all of the small display screens on the right-hand side of the computer screen by mouse clicking on the X box in the upper right-hand corner of each display screen. On the far left-hand side of the screen, midway down, you will see a window labeled Display. **Double left-hand mouse click on the Graph line.** A new window should open. **Scroll down to Fluid Volume and mouse click OK.** A graph of Fluid volume (y-axis) versus time (x-axis) should now appear on the right-hand side of the screen. The data that is plotted on the y-axis can be changed **by moving the** mouse cursor over the label **Fluid Volume** – a box with vertical lines (≡) should appear. **Left-click on the mouse** – a window with several options should appear. **Scroll down to pH and mouse click.** The y-axis should now be set for pH. The x-axis can be changed in similar fashion **by moving the mouse cursor over the label Time (s).** A box with several vertical lines (≡) should appear. **Left-click on the mouse** – a window with several options should appear. **Scroll down to Fluid Volume and mouse click.** The x-axis should now be changed to volume.

EXPERIMENTAL PROCEDURE FOR TITRATION OF ACETIC ACID WITH SODIUM HYDROXIDE

Once the Pasco system has been set up as described in the preceding section, you should be ready to start preparing the first solution for titration. Drain the deionized water that was used in the Drop Counter calibration from the plastic syringe buret. Now fill the syringe buret with 0.10 Molar sodium hydroxide solution. Let a little bit of the NaOH drip from the buret into a waste beaker by opening the bottom stopcock. This will remove any large air pockets from the system. **DO NOT TURN THE TOP STOPCOCK AS THIS WILL AFFECT THE DROP COUNTER CALIBRATION.**

Transfer by pipet 10.0 ml of the commercial vinegar sample into a 400-ml beaker and dilute with about 80 ml of deionized water. The amount of deionized water is not critical. The deionized water is added only to give a convenient so as to cover the glass membrane of the pH electrode. Put the 400-ml beaker with solution under the syringe buret so that the titrant will drip into the titration beaker. (See Figures 15.3 and 15.4 for the apparatus set up.) Insert the pH probe/electrode into the solution, and start the titration by **mouse clicking on the Start button** at the same time that you open the bottom stopcock all of the way. Throughout the titration continuously stir the contents with a glass stirring rod or coffee stirrer (magnetic stirrer and stirring bar if available). You may have to add more sodium hydroxide to the syringe in order to maintain a constant dripping rate and drop size. Initially you should observe a gradual, slow increase in the measured pH. Once you get near the equivalence point, the pH should increase sharply and then level out once the sodium hydroxide is in excess. Go past the equivalence a couple of two or three ml so that you can observe the entire titration curve. Stop the titration by closing the bottom stopcock at the same time that you **mouse click on the Stop button.** Directions for changing the x-axis and y-axis scales are found in Experiment 8 and in Appendix A at the back of the lab manual. Numerical values of pH at specific points along the titration curve can be obtained by **mouse clicking on the Smart Tool on the graph toolbar** (sixth box from left-hand side). An "+ - grid" appears, and one can move the grid to specific parts of the titration curve by positioning the mouse cursor in the center of the grid, holding down on the mouse button, and then dragging the cursor to whatever part of the graph you want to examine. Record the numerical value of the pH of the solution halfway between the starting point and the equivalence point (see Figure 15.1).

If you want to view the first derivative curve you need to have the pH titration curve displayed on the computer screen. **Mouse click on the Calculator button** (first button to the right of Fit ▼) on the Graph screen. You now need to define the mathematical function you want to calculate and display. In the definition box type y = . (Remove anything else that might be in the box.) **Mouse click on the Special ▼ button.** A list of options will appear. **Scroll down to derivative (2,x) and mouse click**. In the definition box there should now be y = derivative(2,x). Now you need to tell the software what data to take the first derivative of. **Click on the ▼ to the left of x = pH versus Fluid Volume** in the Variables box. A new screen should appear. **Scroll down to Data Measurement, and mouse click**. A new window appears. **Scroll down to pH versus Fluid Volume and mouse click OK**. Now **mouse click on the √ Accept button**. Close the calculator screen and the first derivative curve should appear on the graph.

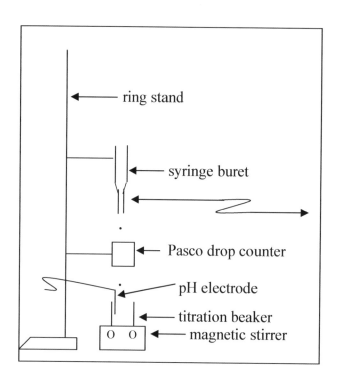

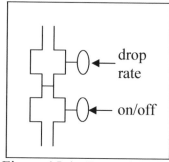

Figure 15.4. Expanded view of syringe tip

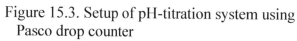

Figure 15.3. Setup of pH-titration system using Pasco drop counter

EXPERIMENTAL PROCEDURE FOR TITRATION OF PHOSPHORIC ACID WITH SODIUM HYDROXIDE

After the Coca Cola has cooled, you can titrate it using the same experimental procedure as was used for the vinegar sample, except that the sodium hydroxide concentration must be 0.010 Molar. The ten-fold dilution is made by transferring 10 ml of 0.100 Molar NaOH by pipet to a beaker or flask, and then adding 90 ml of deionized water to bring the total volume up 100 ml. Replace the 0.10 Molar NaOH that is in the plastic syringe buret with the 0.010 Molar NaOH that you just prepared. Now titrate the entire Coca Cola sample that was prepared. You may want to use a 250-ml beaker for this titration. If you have to add water to get the electrode tip in the solution that is fine. [The H_3PO_4 + NaOH titration curve can be recorded over the CH_3COOH + NaOH titration curve if you want. After the phosphoric acid titration is finished, it is very easy to remove the CH_3COOH + NaOH curve from the computer screen by mouse clicking on the Data button on the Graph top toolbar, and unmarking Run 1. You do not want to exit the system as you may have to recalibrate the Drop Counter.] The pH titration curve has more than one equivalence point because phosphoric acid is a polyprotic acid. When you are sure that the titration is over, mouse click on the Stop button and turn the bottom stopcock to the closed position. Record the required titration data on your laboratory Data Sheet. Do not be concerned if there is not an exact 2:1 mathematical relationship between the volume of NaOH used to reach the second equivalence point versus the volume of NaOH needed to reach the first equivalence. The soft drink may contain a combination of H_3PO_4 and NaH_2PO_4, or may contain a small amount of another acid. If this were the case, then the mathematical 2:1 relationship between endpoints would not hold.

You may want to use the first derivative curve to help in locating the equivalence point volumes. The phosphoric acid pH titration curve is not as sharp as the hydrochloric acid curve. A published article (see J. Murphy, *J. Chem. Educ.*, **60**, 420-421 (1983)) contains an actual pH titration curve for the titration of 25 ml of decarbonated Pepsi-Cola with 0.020 Molar KOH. The author reports that suitable cola beverages include Coca Cola, Pepsi-Cola, Dr. Pepper, Mr. Pibb and a few others. The published article goes on to state that only a small variation between brands was detected, and that the variation was usually no more than the variation between different bottles or cans of the same brand. Manufacturers do periodically change the formulations of soft drinks, and for this reason, no comparison is made between the student-measured values in Experiment 15 and values found in the published literature.

DATA SHEET – EXPERIMENT 15

pH | mls base

Name: _____

Acetic Acid in Vinegar Titration Data

1. Sketch the titration curve of pH (y-axis) versus volume of NaOH added (x-axis):

2. Measured value of pH at equivalence point: _____

3. Measured value of pH half-way to equivalence point: _____

4. Experimental value of K_a for acetic acid: _____

5. Literature value of K_a for acetic acid, from lab manual/textbook: _____

6. Experimental value of pK_a for acetic acid: _____

7. Literature value of pK_a for acetic acid, from lab manual/textbook: _____

8. Discuss how you experimental values compare to the values in the laboratory manual or text book used in the accompanying general chemistry lecture course.

Name: _____

Phosphoric Acid in Coca Cola Titration Data

9. Sketch the titration curve of pH (y-axis) versus volume of NaOH added (x-axis):

10. Measured value of pH at first equivalence point: _____

11. Measured value of pH at second equivalence point: _____

12. Measured value of pH halfway to first equivalence point: _____

13. Measured value of pH halfway between first and second equivalence points: _____

14. Measured value of pH halfway between second and third equivalence points: _____

15. Experimental value of K_{a1} for phosphoric acid: _____

16. Experimental value of K_{a2} for phosphoric acid: _____

17. Experimental value of K_{a3} for phosphoric acid: _____

18. Experimental value of pK_{a1} for phosphoric acid: _____

19. Experimental value of pK_{a2} for phosphoric acid: _____

20. Experimental value of pK_{a3} for phosphoric acid: _____

21. Based on the pH indicators listed in Table 14.1, is there a suitable indicator for the first equivalence point in the phosphoric acid titration, if so, which one: _____

22. Based on the pH indicators listed in Table 14.1, is there a suitable indicator for the second equivalence point in the phosphoric acid titration, if so, which one: _____

190

EXPERIMENT 16A: DETERMINATION OF ACETIC ACID IN VINEGAR BY USE OF pH TITRATION CURVE

INTRODUCTION

In today's laboratory experiment you will repeat the determination of acetic acid in vinegar by sodium hydroxide titration:

$$CH_3COOH_{(aq)} + NaOH_{(aq)} \text{ ------> } H_2O + Na^+_{(aq)} + CH_3COO^-_{(aq)}$$

$$\text{moles of acetic acid } = \text{ moles of sodium hydroxide} \qquad (16A.1)$$

$$\text{Molarity of } CH_3COOH \times \text{Volume of } CH_3COOH = \text{Molarity of NaOH} \times \text{Volume NaOH}$$

$$(16A.2)$$

using a pH electrode, rather than a visual indicator, to locate the equivalence point (neutralization point). An advantage of pH electrode over a visual indicator is that the solution does not have to be clear. Think for the moment of how difficult it would have been to observe a color change in the titration of phosphoric acid in Coca Cola.

The glass membrane pH electrode is a type of specific-ion electrode, in that the electrode responds specifically to H_3O^+ ions. The electrode's selectively is due to the special composition of the thin glass membrane that separates an internal reference H_3O^+ solution (inside the electrode) from the solution whose pH one wishes to measure. The membrane is specific in its response toward hydrogen ions up to a pH of about 9. At higher pH values, however, the glass membrane becomes somewhat responsive to sodium ion, as well as other singly charged cations. A detailed understanding of how the electrode functions, and why the membrane is specific H_3O^+ and not other ions, is beyond the scope of today's laboratory experiment, and would require a fundamental understanding of electrochemistry that is more in depth than what is typically covered in introductory general chemistry. All that you need to know for today's experiment is that the electrode measures the pH of the solution, and that the experimental error/uncertainty is larger at the higher pH-values.

In the titration of acetic acid (beaker) with sodium hydroxide (buret), the solution is initial acidic. As sodium hydroxide is added, the concentration of acetic acid decreases and the solution becomes less acidic. At the very beginning of the titration the decrease in acetic acid is not very noticeable as there is quite a bit of acetic acid in vinegar, and each ml of titrant contains a fixed, fairly small number of moles of sodium hydroxide (the number of moles of NaOH would be 0.0001 for 1.0 ml of 0.100 Molar NaOH). For 50.0 mls of let's say 0.100 Molar CH_3COOH, one ml of 0.100 Molar NaOH would reduce the concentration of CH_3COOH to 0.0961 Molar, which is really not much of a change. The pH of the solution would increase slightly. As more and more of the acetic acid has reacted, a one ml addition of sodium hydroxide still makes only a slight difference in the measured pH. For most of this time, one is in the "buffered" region of the titration curve. The largest difference occurs near the equivalence point, when two or three drops of titrant result in a charge increase in the solution's pH. Once an excess of titrant is present, the incremental addition of still more titrant does not significantly change the solution's pH. A graph of the pH of the solution versus the volume of titrant added gives an S-shaped curve. The essentially flat portion of the curve is the "buffer" region of the titration curve. The equivalence point corresponds to the inflection point in the sharply increasing pH portion of the titration curve.

To calculate an approximate value of what the pH should be at the equivalence point for today's titration, a concentration of CH_3COO^- is needed. Since the titration has not yet been performed, the estimation will have to be based on results from a previous year. When I performed the titration of acetic acid in vinegar a couple of years ago, a value of $[CH_3COOH] = 0.749$ Molar was obtained. It should take somewhere around 37.45 ml of 0.100 Molar NaOH to titrate 5 ml of vinegar. Assuming that one placed the 5 ml of vinegar into 45 ml of deionized water, the total volume of the solution should be 82.45 ml. At the equivalence point the solution would be that of a 0.0454 Molar sodium acetate solution. Acetate ion is the conjugate base of the acetic acid, and the acetate ion reacts with water to form

$$CH_3COO^-_{(aq)} + H_2O \quad <\text{------}> \quad CH_3COOH_{(aq)} + OH^-_{(aq)}$$

$$K_b = 1\times10^{-14}/K_a = \frac{[CH_3COOH][OH^-]}{[CH_3COO^-]} \qquad (16A.3)$$

hydroxide ion and molecular acetic acid. For an acid-base conjugate pair, $K_a \times K_b = 1 \times 10^{-14}$. The acid dissociation constant for acetic acid is $K_a = 1.75 \times 10^{-5}$. To calculate the pH of the solution, a three-line equilibrium table is constructed:

	$CH_3COO^-_{(aq)} + H_2O$	$<\text{------}>$	$CH_3COOH_{(aq)} +$	$OH^-_{(aq)}$
initial:	0.0454		0	small
reacts:	-x		x	x
at equilibrium:	0.0454 − x		x	x

The hydroxide ion concentration is calculated by substituting the last row of the table into Eqn. 16.3. The calculated value of $[OH^-] = 5.09 \times 10^{-6}$ is then substituted into $K_w = 1 \times$

$10^{-14} = [H_3O^+][OH^-]$ to give a hydronium ion concentration of $[H_3O^+] = 1.96 \times 10^{-9}$. The pH of the solution is pH = 8.71.

SET UP THE PASCO SYSTEM FOR A pH TITRATION

Enter the Pasco system by **mouse clicking on the DataStudio icon** on the computer screen that comes up when the computer is first turned on. When the screen stating "How would you like to use DataStudio?" comes up, **mouse click on Create Experiment**. Connect the pH probe and Drop Counter accessory to the PowerLink unit. The system should recognize what accessories have been attached. Go to the top tool bar and **mouse click on the Setup button**. The setup window will now open up on the screen. The set up window will be for calibrating either the pH probe or the drop counter. One can go back and forth between the two calibrations by clicking on the ▼ or ▶ symbol to the left of the accessory name. On the screen for the pH/ISE/Temperature sensor, there will be three options displayed on the screen. You will want to **click on the Calibrate button directly across from the pH box**. The pH calibration is a two-point calibration. Place the pH electrode in the first pH standard/buffered solution, **type in the pH of the buffer in the box for Point 1**, wait a few minutes for the pH electrode to reach equilibrium with the solution, and then **mouse click on the Set button**. The first pH point is now set. Take the pH electrode out of the standard solution, hold the electrode over an empty beaker, rinse the electrode with deionized water from your wash bottle, and carefully blot dry with a paper towel, before placing in the second pH standard/buffer solution. Although the pH of a buffered solution is not suppose to change upon addition of a small amount of acid or base, it is good practice to rinse the pH electrode with deionized before placing it into the next solution. The pH of Point 2 is set by placing the pH electrode into the second standard/buffered solution, **typing in the pH in the box for Point 2**, waiting a few minutes for the electrode to reach equilibrium with the buffered solution, and then **mouse clicking on the Set button**. **Click the OK button** when both points have been set.

Next you need to setup the Pas*Port* drop counter and plastic syringe buret following the instructions given in Experiment 15. A steady drop rate of 3 to 4 drops per second is fine. Now you need to tell the system how large the falling drops are. Position the syringe buret over the rectangular opening in the Drop Counter. Leave room under the drop counter to place the titration beaker and a graduated cylinder. Place a graduated cylinder under the Drop Counter, aligned with the buret so that the dripping titrant will fall into the graduated cyclinder. The cylinder will be used to measure the volume of the titrant that is delivered from the buret. **Mouse click on the Setup button** on the top toolbar. The setup window will now reappear on the screen. **Mouse click on the ▼ or**

► **symbol** to the left Drop Counter. The calibrate screen should now appear on the screen. Open the bottom stopcock to allow the liquid drops to fall directly into the graduated cylinder. You do not want the liquid to drops to hit the Drop Counter or to miss the beaker. If for some reason you need to turn off the buret before the calibration is finished, or if the drops miss the buret, turn the bottom stopcock to the horizontal off position. You will need to empty any liquid from the graduated cylinder, close the calibration screen, and then re-enter the Drop Counter calibration screen by mouse clicking on the Setup button on the top tool bar. Once the liquid in the graduated cylinder reaches the 10-ml mark, turn the bottom stopcock to the full horizontal off position. You now need to tell the system how much liquid actually flowed between the Drop Counter cells. Since 10-ml of liquid was collected in the graduated cylinder, **type 10.00 in the box on the calibration screen. Mouse click on the Set button on the calibration screen.** The number that is above the box should change to 10.00 (*e.g.*, the number that you typed). When this has been done, **mouse click OK.** The Drop Counter should now be calibrated. The green light on the Drop Counter should be flashing the entire time the drops are falling. If the green light stays on (does not flash), this indicates that some liquid has splashed onto the lens. Should this happen gently position a paper towel into the rectangular box, and move along the inside walls to remove the splashed water.

After the pH electrode and Drop Counter have been calibrated, you will need to set the system up to display the measured data as a pH titration curve, *i.e.*, pH (y-axis) versus volume of titrant (x-axis) plot. Close all of the small display screens on the right-hand side of the computer screen by mouse clicking on the X box in the upper right-hand corner of each display screen. On the far left-hand side of the screen, midway down, you will see a window labeled Display. **Double left-hand mouse click on the Graph line.** A new window should open. **Scroll down to Fluid Volume and mouse click OK.** A graph of Fluid volume (y-axis) versus time (x-axis) should now appear on the right-hand side of the screen. The data that is plotted on the y-axis can be changed **by moving the mouse cursor over the label Fluid Volume** – a box with vertical lines (≡) should appear. **Left-click on the mouse** – a window with several options should appear. **Scroll down to pH and mouse click.** The y-axis should now be set for pH. The x-axis can be changed in similar fashion **by moving the mouse cursor over the label Time (s).** A box with several vertical lines (≡) should appear. **Left-click on the mouse** – a window with several options should appear. **Scroll down to Fluid Volume and mouse click.** The x-axis should now be changed to volume.

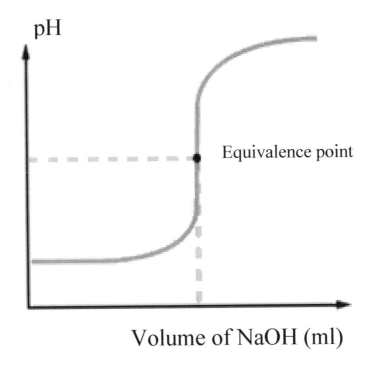

Figure 16A.1. Titration curve for the titration of acetic acid in vinegar with sodium hydroxide.

TITRATION PROCEDURE

Add exactly 10.00 ml of vinegar (pipette) into a 400-ml beaker, and add about 80 ml of deionized water. If you need to add more deionized to cover the pH electrode tip that is fine. The amount of deionized water is not critical. Titrate as you did in Experiment 15. Be sure the addition rate is constant – as one lab partner is stirring the beaker the other lab partner should be continuously filling the plastic syringe burette with standard NaOH solution so as to maintain a constant level. After the equivalence point is reached you can stop the titration. Record the molarity of the sodium hydroxide that is given on the reagent bottle and the volume of sodium hydroxide that it took to reach the equivalence point on your Data Sheet. (The volume of sodium hydroxide is the volume that corresponds to the inflection point (midpoint) of the S-shaped curve; see Figure 16A.1). If you have difficulty locating the inflection point on the titration curve, you can use the first-derivative curve. Instructions for generating the first derivative curve are found in Experiment 15, and in Appendix A at the back of the laboratory manual. Calculate the molarity of acetic acid in vinegar by substituting the measured values into Eqn. 16A.2. How does this value compare to the amount of acetic acid that is labeled on the bottle? To do this comparison, you need to convert the molarity that you determined into a mass/volume percentage. Mass/volume percentage is defined as 100 times the mass of the solute (in grams) divided by the volume of the solution analyzed (in ml). The "100" converts the value to percent. Molarity is converted to mass/volume percent by:

mass/volume % = 100 × molarity of CH_3COOH × (molar mass of CH_3COOH/1000)

(16A.4)

Compare the relative percent error:

% relative error = 100 × [(experimental value – accepted value)/accepted value]

(16A.5)

from the visual phenolphthalein titration performed back in Experiment 14 to the relative error from today's pH titration curve measurement. Which method gave the best experimental results?

DATA SHEET – EXPERIMENT 16A

vinegar

Name: _____

1. Molarity of sodium hydroxide, M: _____

2. Volume of sodium hydroxide to equivalence point, ml: _____

3. Molarity of acetic acid in vinegar, M: _____

4. % Mass/Volume of acetic acid in vinegar, this experiment: _____

5. % Mass/Volume of acetic acid in vinegar, Experiment 14: _____

6. % Mass/Volume of acetic acid (on label): _____

7. % Relative error in experimental % Mass/Volume of acetic acid in vinegar from this experiment: _____

8. % Relative error in experimental % Mass/Volume of acetic acid in vinegar from Experiment 14: _____

EXPERIMENT 16B: TITRIMETRIC ANALYSIS OF A PHOSPHORIC ACID AND SODIUM DIHYDROGEN PHOSPHATE MIXTURE

INTRODUCTION

Titrimetric analysis can provide valuable information in regards to the chemical composition of solutions containing more than a single acid. Last week you experimentally generated the pH titration curve for the titration of the triprotic phosphoric acid with sodium hydroxide. The titration curve showed two equivalence points corresponding to the removal of the first and second proton from H_3PO_4. The third equivalence point was not observed because the titrant was too dilute to reach the high pH necessary for complete removal of the third proton. Careful examination of the observed titration curve would reveal that twice as many milliliters of sodium hydroxide was required to reach the second equivalence point than to reach the first equivalence point. This mathematical relationship can be quite useful in determining whether additional acids are present in a phosphoric acid solution.

In today's laboratory experiment you will titrate a solution containing both H_3PO_4 and NaH_2PO_4 with sodium hydroxide. The molar concentration of H_3PO_4 and NaH_2PO_4 will be calculated using the volume of sodium hydroxide required to reach the first and second equivalence points. The experiment illustrates how a simple neutralization titration can be employed to analyze mixtures, from a qualitative as well as a quantitative standpoint.

COMPATIBILITY SERIES

Before discussing the experiment method, lets construct the compatibility series for phosphoric acid. The compatibility series is generated by listing phosphoric acid and each subsequent conjugate base in order. At the top of the table are listed the very strong acids, like HCl, HBr, *etc.*, and at the bottom of the table are listed the strong bases, such as NaOH and KOH. At equilibrium, no more than two adjacent lines of chemical species can co-exist in an appreciable amount in any solution. Unidirectional chemical reactions will eliminate the others.

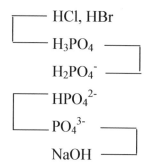

For example, hydrochloric acid and phosphoric acid, H_3PO_4, can co-exist in solution. Phosphoric acid can co-exist with its conjugate base $H_2PO_4^-$. However, if one were to dissolve Na_2HPO_4 into a solution of H_3PO_4, a chemical reaction would occur

$$H_3PO_{4(aq)} + HPO_4^{2-}{}_{(aq)} ----------> 2\ H_2PO_4^-{}_{(aq)}$$

because H_3PO_4 and HPO_4^{2-} are not on adjacent lines in the compatibility table. The reaction would produce $H_2PO_4^-$. When the reaction was over, the solution would contain only $H_2PO_4^-$ if equal amounts of H_3PO_4 and HPO_4^{2-} were mixed, or either H_3PO_4 and $H_2PO_4^-$ or $H_2PO_4^-$ and HPO_4^{2-}, depending on whether HPO_4^{2-} or H_3PO_4 was the limiting reagent. For all practical purposes the limiting reagent would be completely consumed.

The compatibility chart also describes the order in which a strong base like sodium hydroxide reacts with a solution containing more than one acid. Suppose that sodium hydroxide was added to an acid mixture containing both hydrochloric acid and phosphoric acid. Sodium hydroxide would first react with HCl

$$HCl_{(aq)} + NaOH_{(aq)} --------> H_2O + Na^+{}_{(aq)} + Cl^-{}_{(aq)}$$

and once all of the hydrochloric acid was consumed, the added sodium hydroxide would react with phosphoric acid

$$H_3PO_{4(aq)} + NaOH_{(aq)} -------> H_2O + H_2PO_4^-{}_{(aq)} + Na^+{}_{(aq)}$$

$$H_2PO_4^-{}_{(aq)} + NaOH_{(aq)} -------> H_2O + HPO_4^{2-}{}_{(aq)} + Na^+{}_{(aq)}$$

$$HPO_4^{2-}{}_{(aq)} + NaOH_{(aq)} -------> H_2O + PO_4^{3-}{}_{(aq)} + Na^+{}_{(aq)}$$

to remove the first, second and third protons in stepwise fashion. The resulting pH titration curve (see Figure 16B.1) would resemble the phosphoric acid pH titration curve, except that the first part of the titration curve has been extended. It would appear that more base was required to remove the first proton than the second proton. Why? Because sodium hydroxide first had to neutralize the most incompatible acid, which in this example was HCl. Hydrochloric acid is furthest away from NaOH on the compatibility series.

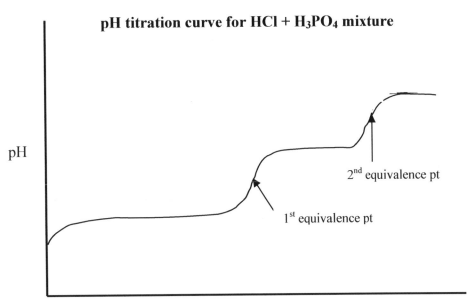

Figure 16B.1. pH titration curve for a mixture of HCl and H_3PO_4 with sodium hydroxide.

Chemical reactions to the first equivalence point

$$HCl_{(aq)} + NaOH_{(aq)} \longrightarrow H_2O + Na^+_{(aq)} + Cl^-_{(aq)}$$

$$H_3PO_{4(aq)} + NaOH_{(aq)} \longrightarrow H_2O + H_2PO_4^-_{(aq)} + Na^+_{(aq)}$$

moles of sodium hydroxide added = moles of hydrochloric acid +

moles of phosphoric acid (16B.1)

molarity of NaOH × volume of NaOH to first equivalence point = moles of HCl

+ moles of H_3PO_4 (16B.2)

and to the second equivalence point

$$HCl_{(aq)} + NaOH_{(aq)} \longrightarrow H_2O + Na^+_{(aq)} + Cl^-_{(aq)}$$

$$H_3PO_{4(aq)} + 2\,NaOH_{(aq)} \longrightarrow 2H_2O + HPO_4^{2-}_{(aq)} + 2Na^+_{(aq)}$$

moles of sodium hydroxide added = moles of hydrochloric acid +

2 × moles of phosphoric acid (16B.3)

molarity of NaOH × volume of NaOH to first equivalence point = moles of HCl

+ 2 × moles of H_3PO_4 (16B.4)

can be used to determine the number of moles of each acid present.

If the acid mixture contained phosphoric acid and sodium dihydrogen phosphate, then the portion of the titration curve between the first and second equivalence points would be elongated. Sodium hydroxide would react with only H_3PO_4 to the first equivalence point

$$H_3PO_{4(aq)} + NaOH_{(aq)} \xrightarrow{\hspace{1cm}} H_2O + H_2PO_4^-{}_{(aq)} + Na^+{}_{(aq)}$$

moles of sodium hydroxide added = moles of phosphoric acid (16B.5)

molarity of NaOH × volume of NaOH to first equivalence point =

moles of H_3PO_4 (16B.6)

and with both H_3PO_4 and $H_2PO_4^-$ to the second equivalence point

$$H_3PO_{4(aq)} + 2\,NaOH_{(aq)} \xrightarrow{\hspace{1cm}} 2\,H_2O + HPO_4^{2-}{}_{(aq)} + 2\,Na^+{}_{(aq)}$$

$$H_2PO_4^-{}_{(aq)} + NaOH_{(aq)} \xrightarrow{\hspace{1cm}} H_2O + HPO_4^{2-}{}_{(aq)} + Na^+{}_{(aq)}$$

moles of sodium hydroxide added = 2 × moles of phosphoric acid +

moles of dihydrogen phosphate (16B.7)

molarity of NaOH × volume of NaOH to second equivalence point =

2 × moles of H_3PO_4 + moles of $H_2PO_4^-$ (16B.8)

By experimentally measuring the volume of sodium hydroxide needed to reach the first and second equivalence points, the number of moles of H_3PO_4 and NaH_2PO_4 can be calculated. Moreover, one can determine whether or not the unknown mixture contained HCl or NaH_2PO_4 by observing which portion of the pH titration was extended. The mixture would not contain all three acids as HCl and NaH_2PO_4 are incompatible with each other.

SET UP THE PASCO SYSTEM FOR A pH TITRATION

Enter the Pasco system by **mouse clicking on the DataStudio icon** on the computer screen that comes up when the computer is first turned on. When the screen stating "How would you like to use DataStudio?" comes up, **mouse click on Create Experiment**. Connect the Pas*Port* pH probe and Pas*Port* Drop Counter accessory to the PowerLink unit. The system should recognize what accessories have been attached. Go to the top tool bar and **mouse click on the Setup button**. The setup window will now open up on the screen. The set up window will be for calibrating either the pH probe or the drop counter. One can go back and forth between the two calibrations by clicking on the ▼ or ▶ symbol to the left of the accessory name. On the screen for the pH/ISE/Temperature sensor, there will be three options displayed on the screen. You will want to **click on the Calibrate button directly across from the pH box**. The pH calibration is a two-point calibration. Place the pH electrode in the first pH standard/buffered solution, **type in the pH of the buffer in the box for Point 1**, wait a few minutes for the pH electrode to reach equilibrium with the solution, and then **mouse click on the Set button**. The first pH point is now set. Take the pH electrode out of the standard solution, hold the electrode over an empty beaker, rinse the electrode with deionized water from your wash bottle, and carefully blot dry with a paper towel, before placing in the second pH standard/buffer solution. Although the pH of a buffered solution is not suppose to change upon addition of a small amount of acid or base, it is good practice to rinse the pH electrode with deionized before placing it into the next solution. The pH of Point 2 is set by placing the pH electrode into the second standard/buffered solution, **typing in the pH in the box for Point 2**, waiting a few minutes for the electrode to reach equilibrium with the buffered solution, and then **mouse clicking on the Set button. Click the OK button** when both points have been set.

Next you need to setup the Pas*Port* drop counter and plastic syringe buret following the instructions given in Experiment 15. A steady drop rate of 3 to 4 drops per second is fine. Now you need to tell the system how large the falling drops are. Position the syringe buret over the rectangular opening in the Drop Counter. Leave room under the drop counter to place the titration beaker and a graduated cylinder. Place a graduated cylinder under the Drop Counter, aligned with the buret so that the dripping titrant will fall into the graduated cyclinder. The cylinder will be used to measure the volume of the titrant that is delivered from the buret. **Mouse click on the Setup button** on the top toolbar. The setup window will now reappear on the screen. **Mouse click on the ▼ or ► symbol** to the left Drop Counter. The calibrate screen should now appear on the screen. Open the bottom stopcock to allow the liquid drops to fall directly into the graduated cylinder. You do not want the liquid to drops to hit the Drop Counter or to miss the beaker. If for some reason you need to turn off the buret before the calibration is finished, or if the drops miss the graduated cylinder, turn the bottom stopcock to the horizontal off position. You will need to empty any liquid from the graduated cylinder, close the calibration screen, and then re-enter the Drop Counter calibration screen by mouse clicking on the Setup button on the top tool bar. Once the liquid in the graduated cylinder reaches the 10-ml mark, turn the bottom stopcock to the horizontal off position. You now need to tell the system how much liquid actually flowed between the Drop Counter cells. Since 10-ml of liquid was collected in the graduated cylinder, **type 10.00 in the box on the calibration screen. Mouse click on the Set button on the calibration screen.** The number that is above the box should change to 10.00 (*e.g.*, the number that you typed). When this has been done, **mouse click OK**. The Drop Counter should now be calibrated. The green light on the Drop Counter should be flashing the entire time the drops are falling. If the green light stays on (does not flash), this indicates that some liquid has splashed onto the lens. Should this happen gently position a paper towel into the rectangular box, and move along the inside walls to remove the splashed water.

After the pH electrode and Drop Counter have been calibrated, you will need to set the system up to display the measured data as a pH titration curve, *i.e.*, pH (y-axis) versus volume of titrant (x-axis) plot. Close all of the small display screens on the right-hand side of the computer screen by mouse clicking on the X box in the upper right-hand corner of each display screen. On the far left-hand side of the screen, midway down, you will see a window labeled Display. **Double left-hand mouse click on the Graph line**. A new window should open. **Scroll down to Fluid Volume and mouse click OK**. A graph of Fluid volume (y-axis) versus time (x-axis) should now appear on the right-hand side of the screen. The data that is plotted on the y-axis can be changed **by moving the mouse cursor over the label Fluid Volume** – a box with vertical lines ($\equiv$) should appear. **Left-click on the mouse** – a window with several options should appear. **Scroll down to pH and mouse click.** The y-axis should now be set for pH. The x-axis can be changed in similar fashion **by moving the mouse cursor over the label Time (s)**. A box with several vertical lines ($\equiv$) should appear. **Left-click on the mouse** – a window with several options should appear. **Scroll down to Fluid Volume and mouse click**. The x-axis should now be changed to volume.

TITRATION PROCEDURE

The unknown solution for today's laboratory experiment will be a mixture of phosphoric acid and sodium dihydrogen phosphate. **[Note to Instructor: The directions given assume that the molar of H_3PO_4 + molarity of $NaH_2PO_4 \approx 0.10$ Molar.]** Add exactly 25.00 ml of unknown solution (pipette) into a 400-ml beaker, and add about 80 ml of deionized water. If you need to add more deionized water to cover the pH electrode tip that is fine. The amount of deionized water is not critical. Titrate with 0.100 Molar sodium hydroxide as you did in Experiment 15. Be sure the addition rate is constant – as one lab partner is stirring the beaker the other lab partner should be continuously filling the plastic syringe burette with standard NaOH solution so as to maintain a constant level. After the second equivalence point is reached you can stop the titration. As a ballpark figure in regards to the volume of sodium hydroxide that you will need to add, if the unknown solution is entirely phosphoric acid, the second equivalence point should be reached when 50 ml of 0.10 Molar sodium hydroxide has been added. If the unknown solution is 0.05 Molar phosphoric acid + 0.05 Molar sodium dihydrogen phosphate, it should take 37.5 ml of 0.10 Molar sodium hydroxide. You will need to go a few ml past the second equivalence point to complete the second S-shape curve. Record the molarity of the sodium hydroxide that is given on the reagent bottle and the volumes of sodium hydroxide that it took to reach the first and second equivalence points on your Data Sheet. If you have difficulty locating the inflection point on the titration curve, you can use the first-derivative curve. Instructions for generating the first derivative curve are found in Experiment 15, and in Appendix A at the back of the laboratory manual. Calculate the number of moles of H_3PO_4 and NaH_2PO_4 in the unknown sample by substituting the molarity of sodium hydroxide and measured equivalence point volumes of sodium hydroxide into Eqns. 16B.6 and 16B.8. The two equations are then solved simultaneously. The stoichiometric molarity of H_3PO_4 and NaH_2PO_4 in the unknown sample, $[H_3PO_4]$ and $[NaH_2PO_4]$, is obtained by dividing the number of moles of each compound by the volume of the unknown sample analyzed, which in the present case was 25 ml.

DATA SHEET – EXPERIMENT 16B

Phosphoric acid

Name: _____

pH Titration Curve Measurement for H$_3$PO$_4$ + NaH$_2$PO$_4$ Mixture

1. Molarity of sodium hydroxide, M: _____

2. Sketch of pH titration curve:

3. Volume of sodium hydroxide to first equivalence point, ml: _____

4. Volume of sodium hydroxide to second equivalence point, ml: _____

5. Moles of phosphoric acid in unknown sample: _____

6. Molarity of phosphoric acid in unknown sample, M: _____

7. Moles of sodium dihydrogen phosphate in unknown sample: _____

8. Molarity of sodium dihydrogen phosphate in unknown sample, M: _____

Post Laboratory Calculation

     25 ml of an unknown solution containing both HCl and H$_3$PO$_4$ was titrated with 0.100 Molar NaOH. It took 25.43 ml of NaOH to reach the first equivalence point, and a total of 43.01 ml of NaOH to reach the second equivalence point.

9. Molarity of HCl in unknown sample, M: _____

10. Molarity of H$_3$PO$_4$ in unknown sample, M: _____

EXPERIMENT 17: DETERMINATION OF pK$_a$ AND MOLAR MASS OF AN UNKNOWN ACID

INTRODUCTION

Thus far during the two-semester general chemistry laboratory sequence you have learned how to identify an unknown substance through its physical and chemical properties (Experiment 2), and have determined the molar mass of a compound by two different experimental methods. In the first of the two molar mass determinations (Experiment 4), a sample of butane gas was trapped. By measuring the pressure, volume, temperature and mass of the trapped butane gas, you calculated the molar mass of butane by substituting the measured quantities into the ideal gas law method. This particular experimental method works fairly well for volatile compounds. The second molar mass determination (Experiment 9) was more for nonvolatile liquids and solids. A known mass of the compound was dissolved in a known quantity of the solvent, and the freezing point temperature of the resulting mixture was measured by the cooling-curve method. The change in the freezing point temperature of the mixture relative to the pure solvent (freezing point depression) is related to the molal concentration of the dissolved solute. By substituting the measured quantities into the freezing point depression equation, the molar mass of the dissolved compound was calculated.

Today's laboratory experiment is going to extend the concepts that you learned previously to identification of an unknown weak organic acid based on measuring the dissociation constant(s) and molar mass of the unknown compound. The experimental method will involve determining the pH titration curve for the titration of the weak acid with a standardized NaOH solution. You already know how to determine the acid dissociation constants from the pH-titration curve. You have used this method already, back in Experiment 15. The molar mass determination is based on the volume of titrant that it takes to reach the equivalence point. For a monoprotic acid like acetic acid, there is only one equivalence point

$$HA_{(aq)} + NaOH_{(aq)} \ \text{-----}> H_2O + Na^+_{(aq)} + A^-_{(aq)}$$
$$\text{moles of HA} = \text{moles of NaOH} \tag{17.1}$$

$$\text{mass of HA/molar mass of HA} = \text{molarity of NaOH} \times \text{volume of NaOH (in liters)} \tag{17.2}$$

which corresponds to the removal of the only transferable proton. The molar mass of an unknown monoprotic acid is calculated by substituting the volume of NaOH that it takes to reach the equivalence point, the mass of the unknown acid used, and the molarity of

NaOH into Eqn. 17.2. The only quantity that is not known is the molar mass of HA, which you want to calculate.

For a diprotic acid (such as carbonic acid or oxalic acid):

$$H_2A_{(aq)} + NaOH_{(aq)} \longrightarrow HA^-_{(aq)} + H_2O + Na^+_{(aq)}$$

moles of H_3A = moles of NaOH $\hspace{4cm}$ (17.3)

mass of H_2A/molar mass of H_2A =

molarity of NaOH × volume of NaOH to first equivalence point (in liters)

$\hspace{12cm}$ (17.4)

$$H_2A_{(aq)} + 2\,NaOH_{(aq)} \longrightarrow A^{2-}_{(aq)} + 2\,H_2O + 2\,Na^+_{(aq)}$$

moles of H_2A = (1/2) × moles of NaOH $\hspace{3cm}$ (17.5)

mass of H_2A/molar mass of H_2A =

(1/2) × molarity of NaOH × volume of NaOH to second equivalence point (liters)

$\hspace{12cm}$ (17.6)

or for a triprotic acid (such as phosphoric acid or citric acid):

$$H_3A_{(aq)} + NaOH_{(aq)} \longrightarrow H_2A^-_{(aq)} + H_2O + Na^+_{(aq)}$$

moles of H_3A = moles of NaOH $\hspace{4cm}$ (17.7)

mass of H_3A/molar mass of H_3A =

molarity of NaOH × volume of NaOH to first equivalence point (in liters)

$\hspace{12cm}$ (17.8)

$$H_3A_{(aq)} + 2\,NaOH_{(aq)} \longrightarrow HA^{2-}_{(aq)} + 2\,H_2O + 2\,Na^+_{(aq)}$$

moles of H_3A = (1/2) × moles of NaOH $\hspace{3cm}$ (17.9)

mass of H_3A/molar mass of H_3A =

(1/2) molarity of NaOH × volume of NaOH to second equivalence point (in liters)

$\hspace{12cm}$ (17.10)

$$H_3A_{(aq)} + 3\,NaOH_{(aq)} \longrightarrow A^{3-}_{(aq)} + 3\,H_2O + 3\,Na^+_{(aq)}$$

moles of H_3A = (1/3) × moles of NaOH $\hspace{3cm}$ (17.11)

mass of H_3A/molar mass of H_3A =

(1/3) molarity of NaOH × volume of NaOH (in liters) to third equivalence point

$\hspace{12cm}$ (17.12)

more than one equivalence point will be observed. The first equivalence point corresponds to the removal of the first transferable proton, the second equivalence point corresponds to the removal of the second transferable proton, and if you happen to have an unknown that is triprotic acid, the third equivalence point (if it is observed) corresponds to the removal of the third transferable proton. The mathematical equations for calculating the molar mass of a diprotic acid (Eqns. 17.4 and 17.6) and for calculating the molar mass of a triprotic acid (Eqns. 17.8, 17.10 and 17.12) are given above for each of the equivalence point volumes. The volumes of sodium hydroxide in the respective equations refer to the total volume of sodium hydroxide that it takes to reach the respective endpoint. If your unknown is a polyprotic acid, calculate the molar mass of the acid using all of the equivalence point volumes of NaOH.

Please note that in the case of triprotic acids, it is sometimes difficult to remove the third H^+. Very often the last equivalence point is not observed, unless one is using a very concentrated solution of NaOH for the titrant. The molarity of the sodium hydroxide that you will be using in today's laboratory experiment is only 0.100 Molar, which may not be concentrated enough to remove the third H^+ from some of the unknown acids. This is generally true for triprotic acids having a third dissociation of $pK_a > 10$. A word of caution – just because you do not observe a third equivalence point, this does not necessarily mean that the unknown acid is not triprotic. Base the identification of the unknown acid on all of the information that you have, the observed pH-titration curve, the numerical value(s) of the acid dissociation constant(s), and the molar mass of the unknown acid. A list of possible unknown acids is given in Table 17.1, along with the pertinent acid dissociation constants and molar masses.

MOLAR MASS AND pK_a OF AN UNKNOWN ACID BASED ON pH TITRATION CURVE MEASUREMENTS

Set up the Pasco system to record a pH titration. Refer back to the instructions given in Experiment 15 for calibrating the pH electrode and drop counter. Once the pH electrode and drop counter are calibrated you are ready to begin preparing the solution for titration. Weigh an empty 250-ml beaker on the electronic top loading balance. Record the mass of the empty beaker on the laboratory Data Sheet. Now add about 0.4 grams of the unknown acid to the beaker and reweigh. (For a liquid unknown, this would be about 0.4 ml to give you an idea of how much to pour into the beaker.) Record the mass of the beaker + unknown acid on the Data Sheet. The mass of the unknown acid analyzed is calculated by difference. The mass of the acid analyzed must be known in order to do the molar mass calculation. Add about 50 ml of deionized water to the 400-ml beaker to dissolve the unknown. Most of the unknowns will dissolve, however, if your does not dissolve readily in water, heat briefly on a hot plate or a Bunsen burner to hand warmness (about 40 – 50 °C). Remove the source of heat. Wait for the solution to come back to about room temperature before titrating. Acid dissociation constants do depend on temperature, and the numerical values that are listed in Table 17.1 are for 25 °C. If your acid is all dissolved after the solution cools back to about room temperature titrate slowly using 0.100 Molar NaOH. Record the molar of the sodium hydroxide on the Data Sheet

Table 17.1. List of Possible Organic Acid Unknowns and their Respective Acid Dissociation Constants and Molar Masses

Organic Acid	pK$_a$'s	Molar Mass	Use/Occurrence
acetic acid	4.75	60.05	vinegar
aspartic acid	1.99, 3.90, 10.02	133.10	one of the 20 most common amino acids
benzoic acid	4.20	122.12	sodium salt is used a food preservative, found in oral hygiene products like mouthwashes
butanoic acid	4.76	88.11	notably found in rancid butter, parmesan cheese and vomit
citric acid	3.13, 4.76, 6.40	192.12	found in citrus fruits like lemons and limes, used in select beverages, found in candies
formic acid	3.75	46.03	found in the sting and bites of ants and bees
fumaric acid	3.03, 4.54	116.06	found in paints and varnishes, sometimes used as a food substitute for tartaric acid in beverages or baking powder
glutamic acid	2.23, 4.42, 9.95	147.13	one of the 20 most common natural amino acids
glycine	2.35, 9.78	75.07	an amino acid
glycolic acid	3.83	76.05	used in skin creams, dermatological skin peels
lactic acid	3.86	90.08	responsible for sour taste of old milk, used in production of dairy products such as cheese and yogart, sometimes used as a food additive in salid dressings
maleic acid	1.92, 6.22	116.07	oil and fat preservative, used as food acidulant
malic acid	3.40, 5.05	134.07	found in apples
malonic acid	2.85, 5.70	104.06	found in aconite, used in the manufacture of barbiturates
oxalic acid	1.25, 4.27	90.03	found in rhubarb, spinach, eggplant, green beans
pyruvic acid	5.49	88.06	created in the body when sugars are metabolized
salicylic acid	2.98	138.12	aspirin, found in many over-the-counter acne medications
succinic acid	4.21, 5.64	118.09	Kreb's citric acid cycle, remedy for alcohol hangover
tartaric acid	3.04, 4.36	150.09	found in grapes and tamarinds, main acid in wines

in the appropriate space. **(If the acid does not all dissolve skip down to the next paragraph.)** Be sure to open the stopcock on the syringe burette at the same time that you click on the Start button. **Click on the Stop button** when the titration is over. You can generally tell when the titration is over by looking at the titration curve, and by noting the pH of the solution as it is recorded. Be sure that you don't stop the titration too soon as you want to observe all of the equivalence points if possible. Once the titration is over, carefully examine the titration curve, and record the volume of sodium hydroxide that it took to reach each observed equivalence point, and the value of the pH at halfway to the first equivalence, and in the case of polyprotic acids, the pH halfway between each two consecutive equivalence points. For example, in the case of a diprotic acid, you would have the pH halfway to the first equivalence point, and the pH halfway between the first and second equivalence points. Record this information on the laboratory Data Sheet. Calculate the molar mass of the unknown acid from the total volume of sodium hydroxide that it took to reach each equivalence point. (See Eqn. 17.2 for a monoprotic acid; Eqns. 17.4 and 17.6 for a diprotic acid; and Eqns. 17.8, 17.10 and 17.12 for a triprotic acid.) Identify the unknown acid by comparing the molar mass and acid dissociation constants that you measured to the values given in Table 17.1.

If your unknown acid does not dissolve, then it will not be possible to do the molar mass determination. The identification of the unknown will have to be based solely on the value(s) of the experimentally determined acid dissociation constant(s). If the sample does not completely dissolve, separate the filtrate (liquid) from the undissolved solid by either decantation or filtration. Once the undissolved residue has been removed, titrate slowly using 0.100 Molar NaOH. (Your sample may not contain very much acid!!) Be sure to open the stopcock on the syringe burette at the time that you click on the Start button. **Click on the Stop button** when the titration is over. You can generally tell when the titration is over by looking at the titration curve, and by noting the pH of the solution as it is recorded. Be sure that you don't stop the titration too soon as you want to observe all of the equivalence points if possible. Once the titration is over, carefully examine the titration curve, and record the value of the pH at halfway to the first equivalence, and in the case of polyprotic acids, the pH halfway between each two consecutive equivalence points. For example, in the case of a diprotic acid, you would have the pH halfway to the first equivalence point, and the pH halfway between the first and second equivalence points. Record this information on the laboratory Data Sheet. Identify the unknown acid by comparing the acid dissociation constants that you measured to the values given in Table 17.1.

DATA SHEET – EXPERIMENT 17

pH

mls of base

Name: _____

Molar Mass and Acid Dissociation Constant Determination

1. Mass of empty beaker, g: _____

2. Mass of beaker + unknown acid, g: _____

3. Mass of unknown acid, g: _____

4. Did the acid all dissolve, yes or no: _____

5. Molarity of the NaOH solution, M: _____

5. Volumes of NaOH at the equivalence points, ml: _____

   [however many apply] _____

   _____

6. Experimental pK 's determined from graphs: pK_{a1} = _____

   [however many apply] pK_{a2} = _____

   pK_{a3} = _____

7. Calculated molar masses based on equivalence points,
   Value from first equivalence point, grams/mole: _____

   [however many apply] _____

   _____

   [Note: If the unknown did not dissolve, write did not dissolve in
   the boxes for above]

8. Identity of unknown acid: _____

EXPERIMENT 18: ACIDS AND BASES IN COMMON HOUSEHOLD PRODUCTS AND pH MEASUREMENTS FOR CAREFULLY PREPARED BUFFERED SOLUTIONS

INTRODUCTION

In today's laboratory experiment you will make pH measurements to determine whether various common household products and carefully prepared chemical solutions are acidic or basic. Water undergoes a self-dissociation reaction:

$$H_2O_{(liq)} + H_2O_{(liq)} \quad <------> \quad H_3O^+_{(aq)} + OH^-_{(aq)}$$

in which one water molecule serves as an hydrogen ion acceptor (base), while the other water molecule is an hydrogen ion donor (acid). Sometimes the equilibrium will be written in short-hand form:

$$H_2O_{(liq)} \quad <------> \quad H^+_{(aq)} + OH^-_{(aq)}$$

Hydronium and hydrogen ions are used interchangeable in these chemical reactions.

As in the case of any equilibrium reaction, the equilibrium constant is written as

$$K_{eq} = K_w = 1 \times 10^{-14} = [H_3O^+][OH^-] \tag{18.1}$$

the product of the molar concentrations of the products raised to the power of the coefficients in the balanced chemical reaction divided by the molar concentration product of the reactants, also raised to the power of the coefficients in the balanced chemical reaction. Solvents do not appear in the equilibrium constant expression. The first equilibrium is formally more correct in that the isolated proton (*e.g.*, hydrogen ion) does not exist per se in water. The proton is coordinated to a water molecule through one of the lone electron pairs on the oxygen atom. At 25 °C the measured value of the ionization constant is $K_w = 1 \times 10^{-14}$. The very small numerical value indicates that the equilibrium is very product-favored.

Equation 18.1 applies to both pure water and to all aqueous solutions. In deionized water the only source of hydronium (hydrogen) ions is from the dissociation reaction, hence $[H_3O^+] = [OH^-]$. Substitution of the mathematical constraint into Eqn. 18.1 gives a value of 1×10^{-7} Molar for both $[H_3O^+]$ and $[OH^-]$. The pH of the solution is defined as

$$pH = - \log [H_3O^+] \tag{18.2}$$

the negative logarithm of the hydronium ion molar concentration. The relative concentrations of H_3O^+ and OH^- ions indicate the acidic, neutral or basic nature of the aqueous solution. For all aqueous solutions the following three possibilities apply:

acidic solution:	$[H_3O^+] > 1.0 \times 10^{-7}$	$[OH^-] < 1.0 \times 10^{-7}$
	pH < 7.0	pOH > 7.0
neutral solution:	$[H_3O^+] = 1.0 \times 10^{-7}$	$[OH^-] = 1.0 \times 10^{-7}$
	pH = 7.0	pOH = 7.0
basic solution:	$[H_3O^+] < 1.0 \times 10^{-7}$	$[OH^-] > 1.0 \times 10^{-7}$
	pH > 7.0	pOH < 7.0

The H_3O^+ and OH^- ion concentrations must always multiply to equal 1.0×10^{-14}, and the pH and pOH (defined as pH = - log [OH^-]) must always sum to 14.0 at T = 25 °C. These requirements are maintained through the water dissociation equilibrium. If either an acid or base is added to a neutral solution, the autoionization equilibrium reaction involving H_3O^+ and OH^- will be disturbed. A product has been added to a system at equilibrium. According to Le Chatelier's Principle, the equilibrium shifts in such as way as to offset the effect of any disturbance. The system compensates for the addition of a product by consuming part of the product that was just added. In consuming part of the added H_3O^+/OH^-, an equal amount of the OH^-/H_3O^+ initially present was also consumed in the process. When may wonder why the addition of H_2O to the system does not increase the H_3O^+ and OH^- ion concentrations. The answer is quite simple. For Le Chatelier's Principle the added (or removed) chemical must appear in the equilibrium constant expression. Water does not appear in Eqn. 18.1.

pH DETERMINATION ON HOUSEHOLD PRODUCTS

In this part of the experiment you will measure the pH of several common household products, and make an educated guess in regards to what chemicals might be present in the product to cause the observed acid or base behavior. Some the information you may already know as the household product may have been used in an earlier laboratory experiment this semester. For example, one of the household items tested will be vinegar. In Experiments 14 and 16A the concentration of acetic acid in vinegar was determined by titration with sodium hydroxide, using either phenolphthalein, methyl orange or a glass pH electrode to signal the equivalence point (neutralization point). The pH of Coca Cola will also be measured. What weak acids do you think might be present in Coca Cola? Well in Experiment 15 you titrated the phosphoric acid in Coca Cola. The carbon dioxide that was initially present in Coca Cola was removed by heating. Milk is another of the household products that will be studied. One of the possible unknown acids in Experiment 17 was lactic acid. Look back at the list of possible unknown acids from Experiment 17.

Not all of the answers will be found in the laboratory manual. That would be too easy. Part of the educational process is learning where to look for needed information. In life answers are not always given to you. One place to look for answers for this experiment would be in the textbook used in your general chemistry lecture course. Bleach (sodium hypochlorite, NaOCl), oven cleaners (sodium hydroxide mixed with detergents), baking sodas and baking powders (mixtures of sodium dihydrogen phosphate and sodium bicarbonate), milk of magnesia (magnesium hydroxide) and lemon juice (citric acid) are examples typically found in introductory chemistry level textbooks.

The labels on the bottles of the household product are also a valuable source of information. As you read the list of major ingredients, look for words like "acid", or words ending with "......amine". And finally do not forget the library and internet. With today's modern search engines, one can search the internet by simply typing in a few key words or phrases. One word of caution though about internet searching. DO NOT BELIEVE EVERYTHING THAT YOU FIND ON THE INTERNET. It is very easy for one to put information on the internet. Much of what is found on the internet has not been peer-reviewed or checked for accuracy.

EXPERIMENTAL PROCEDURE – pH OF HOUSEHOLD CHEMICALS

Now set up the Pasco system to measure pH and calibrate the glass pH electrode (probe) with the standard solution(s) that is (are) provided for you to use. The instructions for calibrating the pH electrode is found in Experiment 15. Measure the pHs of the various household substances, and record the information on the laboratory Data Sheet in the section labeled "pH Determinations on Common Household Products". Please note that not all of the listed household products have been provided. The different laboratory sections are doing different sets of household substances selected from the master list.

pH DETERMINATIONS ON CAREFULLY PREPARED CHEMICAL SOLUTIONS OF KNOWN CONCENTRATION

The second part of today's laboratory experiment will involve measuring the pH of 8 carefully prepared solutions of known chemical concentrations, and then comparing the measured values to values that are theoretically calculated based on the concentration of the acid/base and its dissociation characteristics. Detailed calculations for 4 of the 8 theoretical calculations are provided below. You will need to carefully study the 4 given examples, and then calculate the pH of the remaining four solutions in similar fashion. In treating solutions containing more than a single chemical substance, a hierarchy is used to decide which chemical substance to consider first. The hierarchy is that unidirectional reactions are considered first, followed by equilibrium reactions. A unidirectional reaction is one that, for all practical purposes, goes to completion. Examples of unidirectional reactions include: dissociation of strong electrodes; dissociation of strong acids; and dissociation of strong bases.

The first solution to be considered is 0.100 Molar hydrochloric acid. Hydrochloric acid is a very strong acid

$$HCl_{(aq)} + H_2O \longrightarrow H_3O^+_{(aq)} + Cl^-_{(aq)}$$

which completely dissociates in solution. If one started with 0.100 Molar HCl, then after dissociation of HCl, there would be a concentration of hydronium ion in solution of $[H_3O^+] = 0.100$. The pH $= -\log [H_3O^+] = -\log 0.100 = 1.00$. The pH of sodium hydroxide (strong base) is calculated in similar fashion, except that the concentration of hydroxide ion is known from the complete dissociation reaction, and the hydronium ion concentration is obtained by substituting [OH⁻] into Eqn. 18.1.

The second solution that is considered is 0.100 Molar acetic acid. Acetic acid is a weak acid and its dissociation in water is described by the following chemical equilibrium:

$$CH_3COOH_{(aq)} + H_2O \longleftrightarrow H_3O^+_{(aq)} + CH_3COO^-_{(aq)}$$

$$K_a = 1.75 \times 10^{-5} = \frac{[H_3O^+][CH_3COO^-]}{[CH_3COOH]} \tag{18.3}$$

The dissociation is not complete. The acid-dissociation equilibrium constant, K_a, is subscripted with an "a" to denote that the substance is an acid. Had the equilibrium been for the dissociation of a weak base, then the equilibrium constant would be K_b. The molar concentration of undissociated CH_3COOH at equilibrium is fairly significant. To calculate the pH of solutions containing weak acids (or weak bases) a three-line equilibrium table is constructed:

	CH₃COOH(aq) + H₂O	<------>	H₃O⁺(aq) + CH₃COO⁻(aq)	
initial:	0.100	↑	small	0
reacts:	-x		+ x	+ x
at equilibrium:	0.100 − x	↓	x	x

The first line of the table contains the initial concentrations of the species that are in the solution before the dissociation of the weak acid (or base) is considered. The water column has an arrow drawn through the column because the concentration is not needed in the pH calculation. Water is the solvent and its concentration does not appear in the K_a expression. Initially before any of the acetic acid dissociation, the hydronium ion concentration is the very small amount that comes from the autodissociation of water. Once some of the acetic acid dissolves, the amount of hydronium ion from the water is reduced. Refer to the Le Chatelier's Principle discussion given in the Introduction section. The hydronium ion from the autodissociation of water is negligible. The amount of acetic acid that dissociates is not known, hence "x" appears in three of the columns in the second row. The stoichiometric coefficients in the balanced chemical equilibrium indicate that the dissociation of one acetic acid molecule results in the formation of one hydronium ion and one hydroxide ion. The third row in the table corresponds to the concentrations that are in the solution at equilibrium. To get the numerical values for the ion concentrations, the list row of concentrations are substituted into Eqn. 18.3

$$K_a = 1.75 \times 10^{-5} = \frac{x \cdot x}{0.100 - x} \tag{18.4}$$

and the resulting expression is solved for x. In solving equilibrium constant expressions you will soon discover that one can often neglect $\pm$ x with respect to lead numerical values whenever the equilibrium constant is on the order of 10^{-5} or less, as is the case here. Neglecting "-x" in the denominator, the equation simplifies significantly to $1.75 \times 10^{-5} = x^2/0.100$, which when solved for x gives x = $[H_3O^+]$ = 1.323×10^{-3}. The pH of the solution is then calculated by taking the logarithm of x, and changing the negative sign to positive, $e.g.$, pH = - log 1.323×10^{-3} = 2.88. The pH of the 0.100 Molar ammonia solution is calculated in similar fashion, except that one must work in terms of OH^- ion, ammonia is a weak base ($NH_{3(aq)}$ + H_2O <------> $NH_4^+{}_{(aq)}$ + $OH^-{}_{(aq)}$), having a base-dissociation constant of $K_b = 1.8 \times 10^{-5}$. Once $[OH^-]$ is calculated by solving the equilibrium constant expression (use a three-line table to get the quantities that need to be substituted into the K_b-expression), the value of $[H_3O^+]$ is obtained through Eqn. 18.1.

The next solution, 0.100 Molar sodium acetate, is a bit more complicated. First the strong electrolyte dissociates in water to give

$$NaCH_3COO_{(aq)} \quad \text{--------->} \quad Na^+{}_{(aq)} + CH_3COO^-{}_{(aq)}$$

A solution that is 0.100 Molar sodium ion and 0.100 Molar acetate ion. Acetate ion is the conjugate base of the acetic acid, and the acetate ion reacts with water to form

$$CH_3COO^-{}_{(aq)} + H_2O \quad \text{<------>} \quad CH_3COOH_{(aq)} + OH^-{}_{(aq)}$$

$$K_b = 1 \times 10^{-14} / K_a = \frac{[CH_3COOH][OH^-]}{[CH_3COO^-]} \tag{18.5}$$

hydroxide ion and molecular acetic acid. For an acid-base conjugate pair, $K_a \times K_b = 1 \times 10^{-14}$. To calculate the pH of the solution, a three-line equilibrium table is constructed:

$$CH_3COO^-_{(aq)} + H_2O \longleftrightarrow CH_3COOH_{(aq)} + OH^-_{(aq)}$$

	$CH_3COO^-_{(aq)}$		$CH_3COOH_{(aq)}$	$OH^-_{(aq)}$
initial:	0.1		0	small
reacts:	-x		x	x
at equilibrium:	0.1 – x		x	x

The hydroxide ion concentration is calculated by substituting the last row of the table into Eqn. 18.5. The calculated value of $[OH^-] = 7.56 \times 10^{-6}$ is then substituted into Eqn. 18.1 to give a hydronium ion concentration of $[H_3O^+] = 1.323 \times 10^{-9}$. The pH of the solution is pH = 8.88. You may recognize this particular solution as the equivalence point in the acetic acid + sodium hydroxide ion titrations that were performed back in Experiments 14 and 16A. When acetic acid was titrated with sodium hydroxide it was stated that the pH of the solution at the equivalence point was slightly basic because the solution contained acetate ion. That was why phenolphthalein was picked as the indicator for the titration. Now you know how the equivalence point pH value is calculated for the titration that was performed in these two earlier experiments. The molarity of the acetate ion would be different because the concentration of acetic acid in vinegar is not 0.100. The pH of the 0.100 Molar ammonium chloride solution would be calculated in similar fashion, using K_a rather than K_b. For informational purposes, an ammonium chloride solution would be the equivalence point in the $NH_{3(aq)} + HCl_{(aq)} \longrightarrow NH_4Cl_{(aq)}$ titration.

The final pH calculation that will be illustrated is for the 0.100 Molar acetic acid + 0.100 Molar sodium acetate solution. The strong electrolyte sodium acetate is dealt with first since the dissociation is

$$NaCH_3COO_{(aq)} \longrightarrow Na^+_{(aq)} + CH_3COO^-_{(aq)}$$

unidirectional. At this point, the solution now contains 0.100 Molar CH_3COOH + 0.100 Molar CH_3COO^- and 0.100 Molar Na^+; the latter ion is a spectator ion for the rest of the calculations. Now the dissociation of acetic acid is considered.

$$CH_3COOH_{(aq)} + H_2O \longleftrightarrow H_3O^+_{(aq)} + CH_3COO^-_{(aq)}$$

	$CH_3COOH_{(aq)}$		$H_3O^+_{(aq)}$	$CH_3COO^-_{(aq)}$
initial:	0.100		small	0.100
reacts:	-x		+ x	+ x
at equilibrium:	0.100 – x		x	0.100 + x

222

A three-line table is set up as before, except now the initial concentration of the acetate is 0.100 Molar. Where did this acetate ion come from? The dissociation of sodium acetate produced the acetate that is given on the first line of the table. That is why unidirectional reactions are treated first.

Substituting the last line of the table into Eqn. 18.3:

$$K_a = 1.75 \times 10^{-5} = \frac{x(0.100 + x)}{0.100 - x} \tag{18.6}$$

the hydronium ion concentration is calculated $[H_3O^+] = x = 1.75 \times 10^{-5}$. The calculated hydronium ion concentration corresponds to a pH = 4.76. The numerator and denominator can be simplified by remembering that $\pm$ x-values can be ignored whenever the equilibrium constant is on the order of 10^{-5} or less. The pH of the 0.100 Molar ammonia + 0.100 Molar ammonium chloride solution is calculated in similar fashion, except that one works in terms of K_b and hydroxide ions.

EXPERIMENTAL PROCEDURE – pH OF CHEMICAL SOLUTIONS OF KNOWN CONCENTRATION

Measure the pHs of the eight chemical solutions of known concentration, and record the information on the laboratory Data Sheet in the section labeled "pH Determinations on Solutions of Known Acid/Base Concentrations". Calculate what the pH's of the solutions should be based on the chemical composition of the solutions, and the numerical values of dissociation constants of the acid and/or base dissolved in the solution. Four of the theoretical calculations are given in the laboratory discussion. You need to do the calculations for: (a) the 0.100 Molar sodium hydroxide solution; (b) for the 0.100 molar ammonia solution; (c) for the 0.100 Molar ammonium chloride solution; and (d) for the 0.100 molar ammonia + 0.100 molar ammonium chloride solution. Record your calculated values on the laboratory Data Sheet, as well the calculated values for the other four solutions that are contained in the section titled "pH Determinations of Carefully Prepared Chemical Solutions of Known Concentration".

DATA SHEET – EXPERIMENT 18

pH electrode

Name: _____

pH Determinations on Common Household Products

Substance	pH	Acid/Base/Neutral?	Active Ingredient
Lemon juice	____	_____	_____
Grape juice	____	_____	_____
Vinegar	____	_____	_____
Household ammonia	____	_____	_____
Mr. Clean	____	_____	_____
Milk of magnesia	____	_____	_____
7-Up	____	_____	_____
Coca Cola	____	_____	_____
Bleach	____	_____	_____
Mouthwash (Scope/Listerine/Listermint)	____	_____	_____
Orange juice	____	_____	_____
Black coffee	____	_____	_____
Shampoo	____	_____	_____
Baking soda/powder	____	_____	_____
Milk	____	_____	_____
409	____	_____	_____

pH Chart for Common Household Products

Construct a pH chart for the household substances by writing the name of the each of the substances that you studied on the pH chart below.

$[H_3O^+]$	$[OH^-]$	pH		Household substance
1×10^0	1×10^{-14}	0		
1×10^{-1}	1×10^{-13}	1		
1×10^{-2}	1×10^{-12}	2		
1×10^{-3}	1×10^{-11}	3		
1×10^{-4}	1×10^{-10}	4		
1×10^{-5}	1×10^{-9}	5		
1×10^{-6}	1×10^{-8}	6		
1×10^{-7}	1×10^{-7}	7		
1×10^{-8}	1×10^{-6}	8		
1×10^{-9}	1×10^{-5}	9		
1×10^{-10}	1×10^{-4}	10		
1×10^{-11}	1×10^{-3}	11		
1×10^{-12}	1×10^{-2}	12		
1×10^{-13}	1×10^{-1}	13		
1×10^{-14}	1×10^0	14		

pH Determinations on Solutions of Known Acid/Base Concentrations

0.100 Molar hydrochloric acid

Measured pH: _____

Calculated pH: _____

Name: _____

0.100 Molar sodium hydroxide

        Measured pH: _____

        Calculated pH: _____

0.100 Molar acetic acid

        Measured pH: _____

        Calculated pH: _____

0.100 Molar ammonia

        Measured pH: _____

        Calculated pH: _____

0.100 Molar sodium acetate

        Measured pH: _____

        Calculated pH: _____

0.100 Molar ammonium chloride

        Measured pH: _____

        Calculated pH: _____

0.100 Molar acetic acid + 0.100 Molar sodium acetate

        Measured pH: _____

        Calculated pH: _____

0.100 Molar ammonia + 0.100 Molar ammonium chloride

        Measured pH: _____

        Calculated pH: _____

EXPERIMENT 19: INTRODUCTION TO SPECTROMETRY – VERIFICATION OF BEER'S LAW

INTRODUCTION

The absorption of electromagnetic radiation by ions and molecules serves as the basis for numerous analytical methods of analysis, both qualitative and quantitative. In addition, the measured absorption spectrum provides valuable knowledge concerning the chemical formula, molecular structure and the stability of many chemical species. This laboratory experiment and the next two experiments are devoted to spectroscopic methods. In Experiment 20 you will study a chemical reaction by monitoring the intensity of light that passes through the solution, and in Experiment 21 you will determine the numerical value of an equilibrium constant by measuring the amount of light that the product absorbs. Experiment 26 describes the determination of the equilibrium constant for the acid-base indicator bromothymol blue.

In performing spectrometric measurements it is important to know how the intensity of the incident radiation, I_0, and intensity of the transmitted radiation, I, are related to the amounts or concentrations of compounds being analyzed. The mathematical relationship between these quantities for monochromatic radiation is expressed by the Beer-Lambert law:

$$- \log (I/I_0) = a\, b\, C \qquad (19.1)$$

where b is the path length of the sample in the radiation path, C is the molar concentration of the radiation-absorbing species (called a chromophore), and a is the proportionality constant called the absorptivity coefficient. The absorptivity coefficient is wavelength dependent. If b is expressed in centimeter units, then the proportionality constant, a, is called the molar absorptivity coefficient, ε, with units of liters/(mole cm). In most spectrophotometric determinations the path length is held constant, and the intensity of the transmitted radiation decreases exponentially with chromophore concentration. The term $- \log (I/I_0)$ is given the symbol A, and the name absorbance.

The Beer-Lambert law is

$$A = \varepsilon\, b\, C \qquad (19.2)$$

applicable to only dilute solutions. When the concentration of the chromophore exceeds about 0.01 Molar the absorbing species become so crowded that chromophore-chromophore interactions begin to affect the species' interactions with the incident radiation. Nonlinearity then results. A similar effect is sometimes encountered in solvent media containing low chromophore concentrations but very large concentrations of other chemical species, particularly electrolytes. Close proximity of ions to the chromophore alters the molar absorptivity coefficient of the latter by electrostatic interactions; the effect is lessened by dilution. To minimize such deviations, fairly dilute dye solutions will be studied.

Beer's law is a special case of the more general Eqn. 19.2, and examines only the relationship between absorbance and concentration:

$$A = -\log (I/I_0) = \text{constant} \times C \tag{19.3}$$

The path length is held fixed during the measurement, and is incorporated into the proportionality constant, $e.g.$, constant $= \varepsilon\, b$. With the path length held constant, its functional dependency cannot be determined. That will be the case in today's laboratory experiment. The Pasco colorimeter unit is designed to accept only one size of sample cell.

As you study the various dyes today notice how each dye color is related to the analysis wavelength of the light used to irradiate the solution. When a sample absorbs visible light the color that is perceived is the sum of the remaining colors that are either reflected by or transmitted through the object. An opaque object reflects all light, whereas a transparent object transmits the light. If an object absorbs all of the radiation, then none is reflected back (or transmitted) to observe. As a result the object appears black. The particular color that an object appears results from the preferential absorption of select colors. An object that appears orange in color absorbs all but orange light. An orange color is also perceived when all colors of light except blue reach one's eyes. Blue and orange are complementary colors, and the removal of one color causes the object to appear the other color. Listed below is a partial color chart:

Color of light absorbed	Approximate wavelength region	Color observed
Violet	400 – 420 nm	Green
Indigo	420 – 440 nm	Yellow
Blue	440 – 470 nm	Orange
Blue-green	470 – 500 nm	Red
Green	500 – 520 nm	Red-purple
Yellow-green	520 – 550 nm	Violet
Yellow	550 – 595 nm	Violet-blue
Orange	595 – 620 nm	Blue
Red	620 – 680 nm	Blue-Green
Purple	680 – 780 nm	Green

listing the color of light absorbed, an approximate wavelength region for the absorbed light, and the color that the substance appears. The shades of colors and the corresponding wavelength ranges given are only approximate in that color perception varies from one person to the next. The Pasco colorimeter has four colors of visible light that can be used to irradiate a sample: red (660 nm); green (565 nm); blue (468 nm); and orange (610 nm). Based on the preceding discussion you should be able to make an educated guess regarding the colors of the dyes that will be studied in today's laboratory experiment.

As an informational note, the observed visible absorbance spectrum for dye molecules dissolved in liquid solution typically exhibits a fairly broad absorption spectrum (see Figure 19.1). Ideally one would like to select the analysis wavelength to correspond with the absorption maximum as this would enable one to study more dilute solutions. This is not always possible, particularly if one has a limited number of irradiation wavelengths (colors). The Pasco colorimeter has only the four radiation colors (red, green, blue and orange). Even if the optimum wavelength is not obtainable (achievable), absorption measurements can still be performed at other wavelengths where the dye molecule absorbs. One just will not be able to study solutions having the very dilute dye concentrations.

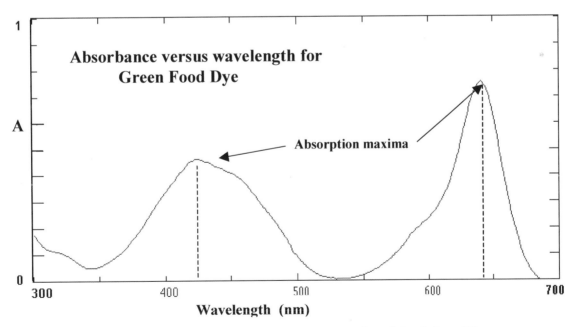

Figure 19.1. Absorption spectrum for a green colored food dye. There are two fairly large absorption maxima. The "best" analysis wavelengths correspond to those at the two absorption maxima, as indicated by the - - - - lines.

231

SETUP AND CALIBRATION OF PASCO COLORIMETER UNIT

Enter the Pasco system by **clicking on the DataStudio icon** on the computer screen. When asked by the system "How would you like to use DataStudio?" **click on Create Experiment**. Connect the Pas*Port* Colorimeter to the PowerLink unit. To zero the instrument fill the glass sample cell with distilled water (or appropriate solvent blank). In Experiment 19 you will use distilled water. Open the top of the sample compartment on the colorimeter unit, insert the sample tube of distilled water, and close the lid completely. You do not want any stray radiation from the room reaching the detector. On the Pas*Port* Colorimeter probe, between the sample holder lid and where the unit is connected into the PowerLink box, you will see a green oval button. **Press the oval green calibrate button**. A green light should illuminate on the oval button to indicate that the calibration is in progress. Wait for the light to turn off and then remove the sample. [Note- a blinking red light on the oval calibration button means either: (a) the dark count is too high; (b) stray light is entering the sample compartment; or (c) the sensor measurement is out of range. The light turns off when the reading is within the normal range. To verify that the instrument is indeed calibrated, reinsert the sample of distilled water, close the lid, and click the Start button on the top tool bar. All color readings should be approximately 100 % transmittance (0 absorbance if the system is set up to display absorbance.)

The Pasco software automatically records the measured intensity data as % Transmittance (if you need to make a conversion to absorbance, the formula is Absorbance = 2 – log % Transmittance). To modify the program to display the measured data as absorbances, **left mouse click on the Setup button on the top toolbar**. Off to the right a new screen titled "Equipment Setup" should appear. Maximize the "Equipment Setup" screen by **mouse clicking on the ☐ box in the upper right-hand corner of the new screen**. When the screen is maximized you should see a series of eight boxes that look as follows:

         ☐ Red (660 nm)   Transmittance %
         ☐ Green (565 nm)   Transmittance %
         ☐ Blue (468 nm)   Transmittance %
         ☐ Orange (610 nm) Transmittance %
         ☐ Red (660 nm)   Absorbance
         ☐ Green (565 nm)   Absorbance
         ☐ Blue (468 nm)   Absorbance
         ☐ Orange (610 nm) Absorbance

An "X" in the box will show what the instrument is currently set up to display. To change the display format, point the mouse cursor in the box and click. Off to the upper left-hand side of the computer screen, under the box titled DataScreen, you should see new types of wavelength readings (color, transmittance or color, absorbance) appearing and disappearing as they changed. For most of applications you will want to display the absorbance as this is the quantity that is related to concentration through Beer's law. Once you have finished changing the display for the wavelengths and way that you want the recorded data displayed, close the PasPort Sensors screen by mouse clicking on the lower of the two X boxes near the upper right-hand corner of the computer screen. **Be careful not to close the DataStudio screen, which is the very top of the two X boxes.**

Now that the system is set up to record and display the intensity data that you want to measure, you need to tell the system whether you want the data to be displayed in a time versus absorbance table format or as a plot of absorbance (y-axis) versus time (x-axis). If you want the data to be displayed in table format, go to the Display section of the left-hand side of the computer screen. **Double left-hand mouse click on the Table line.** A new window will open. Scroll down to the wavelength (color) and type of measurement (Transmittance % or absorbance), and **mouse click OK**. The measured values will be displayed as ordered (x,y) pairs in a time-absorbance table.

If you want the measured intensity data to be displayed in a graphical format of time (x-axis) versus Transmittance % (or Absorbance) (y-axis) **double left-hand mouse click on the Graph button** in the Display section of the left-hand side of the computer screen. Scroll to the wavelength (color) and type of measurement (Transmittance % or absorbance) you want displayed and **mouse click OK**. The system should now be set up to make experimental measurements.

ABSORBANCE MEASUREMENTS ON FOOD DYES

Prepare a colored solution by adding one drop of the yellow food dye to 30 ml of deionized water. Fill the sample cell with this solution, and place the sample cell in the colorimeter sample holder. Be sure to close the lid. Using the set up menu, select the blue light and the absorbance mode for displaying the data (see directions above). Under the display screen **double left-hand mouse click on the Table option for display**, scroll down to blue light absorbance line. **Mouse click OK**. To take the first reading, **mouse click on the Start button on the top tool**, take several seconds of measurements, and then **click on the Stop button**. Record the absorbance reading on the laboratory Data Sheet for the yellow dye. The concentration of this solution will be C = 1.000.

Transfer 10 ml of this solution (by pipet or graduated cylinder) to a clean, 100-ml or 150-ml beaker, and add 10 ml of deionized water (again by pipet or graduated cylinder). Measure the absorbance of the diluted solution, and record the value on the Data Sheet on the line labeled 0.500. The concentration of the diluted solution is only one-half of the concentration of the first solution that you prepared; hence it is labeled 0.500. Repeat the procedure six more times – by drawing off 10 ml of each successive solution, and then diluting with 10 ml of each deionized water. Each successive dilution should half the concentration. The concentrations of the new solutions will be 0.250, 0.125, 0.0625, 0.0312, 0.0156 and 0.00781, respectively. Measure the absorbance of each solution, and record the value on the data sheet.

As you accumulate experimental data from the multiple measurements, it will be necessary to periodically clear the entries from the screen. The Table screen is only so large, and after awhile the columns start to overlap. To clear data in between runs, **mouse click on the Data ▼button** on the Table toolbar. The data that is being displayed in the table will have a √ check-mark to the left of the Run number. Scroll down to the data that you no longer want to display, **mouse click on the run number**, and the √ check-mark should disappear. The runs that you have unmarked should no longer be displayed. The data is still stored, however, and can be easily retrieved by **clicking on the Data ▼button** on the table toolbar. Mark the data that you want redisplayed by placing the cursor of the mouse pointer on the run number and **click OK**. The numbers will be redisplayed in the table.

After you have finished measuring and recording the absorbance data for the yellow dye, you will need to perform a linear least squares analysis of the absorbance (y-axis) versus concentration (x-axis) to see if Beer's law was obeyed over this concentration range. A linear least squares analysis can be accomplished by exiting the current activity, or you can do the linear least squares analysis within the current activity (Experiment button on the top toolbar – New Empty Data Table – described in greater detail in Appendix A.) To do a linear least squares regression, go up to **File button on the top toolbar, click**, and **scroll down to New Activity and mouse click**. There should be no reason to the store the experimental absorbance data as it should be recorded on the Data Sheet. **Click on the Enter Data picture** on the Welcome to DataStudio screen. Enter the concentration (x-value) and absorbance (y-value) data for the yellow dye in the table. Once all of the data has been entered, the linear fit is obtained by **mouse clicking on the Fit button** on the graph toolbar, and **scrolling down to Linear fit**. Record the equation of the straight line and the correlation coefficient on your laboratory Data Sheet. Also record on the laboratory Data Sheet any observations regarding which portion of the graph was linear, and did linearity tail off at either end of the graph. Remember that deviations from the Beer's law linearity are often observed for very dilute, or for very concentrated solutions.

To go back to making absorbance measurements, click on the File button on the top toolbar, mouse click on the new activity line, and this will return you to the Data.

Studio welcome. Click on Create Experiment picture, and you should be back to where you were before. Now perform the same set of experimental measurements for the red food dye (use blue light – solution does absorb green light, but the absorbance is greater for blue light); for the green food dye (solution does absorb red light, however, the absorbance is greatest for orange light), and for the blue food dye (solution absorbance is greatest for orange light). For the each color of dye use the color of light where the absorbance is maximum. Record the experimental values on the Data Sheet under the appropriate dye color. As you finish with each dye, perform the linear least squares regression analysis, and record the pertinent statistical information on the data sheet, as well as any visual observations that you might make concerning whether or not the linear relationship was obeyed over the entire concentration or did it taper off at one or both ends.

MOLAR ABSORPTIVITY COEFFICENT DETERMINATION FOR CRYSTAL VIOLET

In this part of the experiment you will determine the molar absorptivity coefficient, ε, for the dye crystal violet. The molar absorptivity coefficient will be needed in Experiment 19 when you study the decolorization of crystal violet by hydroxide ion. Experiment 19 and today's molar absorptivity constant determination go hand-in-hand. Other dye molecules that can be substituted for crystal violet are malachite green and fuchsin. The three dyes are structurally similar (see Figure 19.2), and each reacts with hydroxide ion.

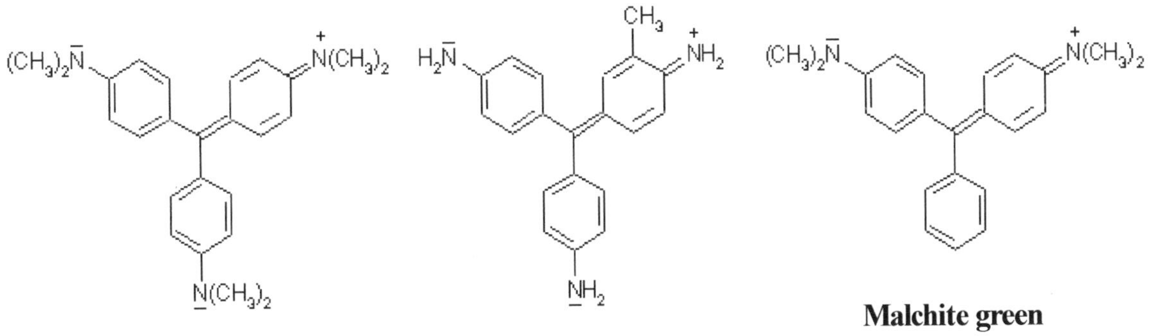

Crystal violet ($\lambda = 590$ nm)   Fuchsin ($\lambda = 545$ nm)   Malchite green ($\lambda = 620$ nm)

Figure 19.2. Molecular structures of the three triphenylmethane dyes: crystal violet; fuchsin and malchite green. The wavelengths for the molar absorptivity coefficient determinations are given in parentheses.

235

The molar absorptivity coefficient determination will be based on crystal violet concentrations of 8.0×10^{-5}, 4.0×10^{-5}, 2.0×10^{-5}, 1.0×10^{-5}, 5.0×10^{-6}, 2.5×10^{-6}, and 1.25×10^{-6} Molar. The stock solution of crystal violet is 8.0×10^{-5} Molar. To prepare the 4.0×10^{-5} Molar solution, you will need to make a two-fold dilution of the stock solution, by transferring by pipet 5 ml of stock solution into a 10-ml volumetric flask. Now dilute to the mark with deionized water. Save a sample of the 8.0×10^{-5} Molar stock solution in a clean, dry 50-ml beaker for the absorbance measurements.

Next pipet 5 ml of the 4.0×10^{-5} Molar crystal violet solution that you just prepared into a second 10-ml volumetric flask. Dilute to the 10-ml mark with deionized water. This will prepare the 2.0×10^{-5} Molar solution. Transfer the remaining 5 mls of the 4.0×10^{-5} Molar crystal violet solution to a clean, dry 50-ml beaker so that you can reuse the first 10-ml volumetric flask to make the next successive dilution. Be sure to carefully label the 50-ml beakers so that you do not mix up the seven crystal violet solutions that you will prepare. In succession, continue to make two-fold dilutions by transferring by pipet the previous solution into a 10-ml volumetric flask, filling the flask to the mark with deionized water, while saving the previous solution in a clean, dry 50-ml labeled beaker. Once you are through with the dilutions you should have seven 50-ml beakers containing: 8.0×10^{-5}, 4.0×10^{-5}, 2.0×10^{-5}, 1.0×10^{-5}, 5.0×10^{-6}, 2.5×10^{-6}, and 1.25×10^{-6} Molar crystal violet.

After the dilutions are complete, you are now ready to measure and record the absorbance of each solution. The wavelength of maximum absorbance for crystal violet is 590 nm (yellow light). The Pasco colorimeter does not come with this particular radiation wavelength. The closest wavelength region that the Pasco system has is orange and green, which fall on either side of yellow in "ROY G BIV". Set the Pasco colorimeter up to measure the absorbance of the Green light by mouse clicking on the Setup button, maximizing the Equipment Setup screen. The screen is maximized, point the mouse cursor to Green (565 nm) Absorbance line and mouse click in the box. Measure the absorbances of the seven crystal violet solutions as you did for the four food dye colors.

Once the absorbances are measured you will need to perform a linear least squares regression analyses of absorbance (y-axis) versus concentration (x-axis). Record the slope of the straight line and the corresponding correlation on the laboratory Data Sheet. Note: In performing the linear regression you may have to remove the data points for the two most concentrated solutions in order to obtain a linear equation. These two solutions are the ones that are most likely to exhibit deviations from Beer's law. Once the slope of the linear equation is obtained, slope = εb, calculate the molar absorptivity by assuming that the path length is b = 1 cm. **Record the value of the molar absorptivity coefficient on the laboratory Data Sheets for both Experiment 19 and Experiment 20. You may need this value for next week's experiment!!**

There is a literature value for the molar absorptivity coefficient of crystal violet. J. C. Turgeon and V. K. LaMer (*J. Am. Chem. Soc.*, **74**, 5988-5995 (1952)) reported a value of 8.5×10^4 liter/(mole cm) for the molar absorptivity coefficient of crystal violet at 590 nm. How does your measured value compare to the published literature value? The values may differ. The molar absorptivity coefficient is wavelength dependent as stated in the laboratory's Introduction. Your value is based on absorbance measurements at 565 nm, whereas the literature value was measured at 590 nm.

NOTE TO INSTRUCTORS – McCormick Assorted Food Colors and Egg Dye were used for the first set of Beer's law measurements. McCormick makes a "Neon" set of assorted food colors and dyes. Wavelength maxima for the second set of dyes are as follows: purple (525 nm); pink (525 nm); blue (626 nm) and green (424 nm large band, 624 nm much smaller band).

DATA SHEET – EXPERIMENT 19

Food Colors & Dyes

Name: _____

Absorbance Measurements on Colored Food Dyes

Yellow Food Dye:

1. Color of analysis light used: _____

2. Measured absorbance for C = 1.000:   _____

3. Measured absorbance for C = 0.500:   _____

4. Measured absorbance for C = 0.250:   _____

5. Measured absorbance for C = 0.125:   _____

6. Measured absorbance for C = 0.0625:   _____

7. Measured absorbance for C = 0.0312:   _____

8. Measured absorbance for C = 0.0156:   _____

9. Measured absorbance for C = 0.000781:   _____

10. Equation for linear least squares regression line: _____

11. Correlation coefficient for line: _____

12. Summary of visual observations regarding linear relationship:

Red Food Dye:

13. Color of analysis light used: _____

14. Measured absorbance for C = 1.000:   _____

15. Measured absorbance for C = 0.500:   _____

16. Measured absorbance for C = 0.250:   _____

17. Measured absorbance for C = 0.125:   _____

18. Measured absorbance for C = 0.0625:   _____

19. Measured absorbance for C = 0.0312:   _____

20. Measured absorbance for C = 0.0156:   _____

21. Measured absorbance for C = 0.000781:   _____

22. Equation for linear least squares regression line: _____

23. Correlation coefficient for line: _____

24. Summary of visual observations regarding linear relationship:

Green Food Dye:

25. Color of analysis light used: _____

26. Measured absorbance for C = 1.000:   _____

27. Measured absorbance for C = 0.500:   _____

28. Measured absorbance for C = 0.250:   _____

29. Measured absorbance for C = 0.125:   _____

30. Measured absorbance for C = 0.0625:   _____

31. Measured absorbance for C = 0.0312:   _____

32. Measured absorbance for C = 0.0156:   _____

33. Measured absorbance for C = 0.000781:   _____

34. Equation for linear least squares regression line: _____

35. Correlation coefficient for line: _____

36. Summary of visual observations regarding linear relationship:

Blue Food Dye:

37. Color of analysis light used: _____

38. Measured absorbance for C = 1.000: _____

39. Measured absorbance for C = 0.500: _____

40. Measured absorbance for C = 0.250: _____

41. Measured absorbance for C = 0.125: _____

42. Measured absorbance for C = 0.0625: _____

43. Measured absorbance for C = 0.0312: _____

44. Measured absorbance for C = 0.0156: _____

45. Measured absorbance for C = 0.000781: _____

46. Equation for linear least squares regression line: _____

47. Correlation coefficient for line: _____

48. Summary of visual observations regarding linear relationship:

Molar Absorptivity Coefficient Determination for Crystal Violet

49. Color of analysis light used: _____

50. Measured absorbance for $C = 8.0 \times 10^{-5}$ Molar: _____

51. Measured absorbance for $C = 4.0 \times 10^{-5}$ Molar: _____

52. Measured absorbance for $C = 2.0 \times 10^{-5}$ Molar: _____

53. Measured absorbance for $C = 1.0 \times 10^{-5}$ Molar: _____

54. Measured absorbance for $C = 5.0 \times 10^{-6}$ Molar: _____

55. Measured absorbance for $C = 2.5 \times 10^{-6}$ Molar: _____

56. Measured absorbance for $C = 1.25 \times 10^{-6}$ Molar: _____

57. Slope of the linear regression of Absorbance vs. Dye concentration: _____

58. Correlation coefficient for linear regression: _____

59. Calculated molar absorptivity coefficient, ε, liter/(mole-cm): _____

60. J. C. Turgeon and V. K. LaMer (*J. Am. Chem. Soc.*, **74**, 5988-5995 (1952)) reported a value of 8.5×10^4 liter/(mole cm) for the molar absorptivity coefficient of crystal violet at 590 nm. How does your measured value compare to the published literature value?

61. There is a valid reason why your value may not be identical to the published literature value. What is the reason? (**Hint**: The reason is stated in the Introduction.)

EXPERIMENT 20: CHEMICAL KINETICS II – DETERMINATION OF THE ORDER OF REACTION AND RATE CONSTANT BASED ON INTEGRAL RATE FORM EXPRESSION

INTRODUCTION

Chemical kinetics is the area of chemistry that is concerned with how fast chemical reactions occur. Some chemical reactions, such as explosions, are complete within fractions of seconds, while other reactions may take thousands or perhaps even millions of years. The formation of diamonds or other minerals in the Earth's crust is an extremely slow chemical process.

The rate of a chemical reaction can be determined by measuring either the rate of disappearance of the reactants or the rate of appearance of a product. In today's experiment the rate law for the reaction for crystal violet with sodium hydroxide according to:

Crystal violet + $OH^-_{(aq)}$ -------> Products

will be determined. Crystal violet is a large organic dye molecule (molar mass of 408 grams/mole). Its intense purple color is due to strong absorption in the yellow-orange region of the visible spectrum, with the absorption maximum at 590 nm. The intense absorption of light is caused by the conjugation (alternating double and single bonds) in the molecule (See Figure 20.1). As the chemical reaction proceeds, the intense purple color slowly disappears. Crystal violet reacts with the hydroxide ions to form a colorless product. From Experiment 19 you learned that the molar concentration of a compound can be determined by measuring its absorbance. Concentration and absorbance are related through the Beer-Lambert law, *e.g.*, $A = \varepsilon\, b\, C$. The spectrophotometric method of analysis that was used in the preceding experiment allows one to follow the concentration of crystal violet as it reacts.

The general rate law expression for the reaction of crystal violet (dye) with hydroxide ion has the form:

$$\text{Rate} = -\,d[\text{Dye}]/dt = k_{rate}\,[\text{Dye}]^x\,[OH^-]^y \qquad (20.1)$$

where $d[\text{Dye}]/dt$ denotes the change in the concentration of dye over time, and k_{rate} is the rate constant. The negative sign in front of $d[\text{Dye}]/dt$ means that the dye concentration decreases with time. The exponents "x" and "y" (called reaction orders) are typically

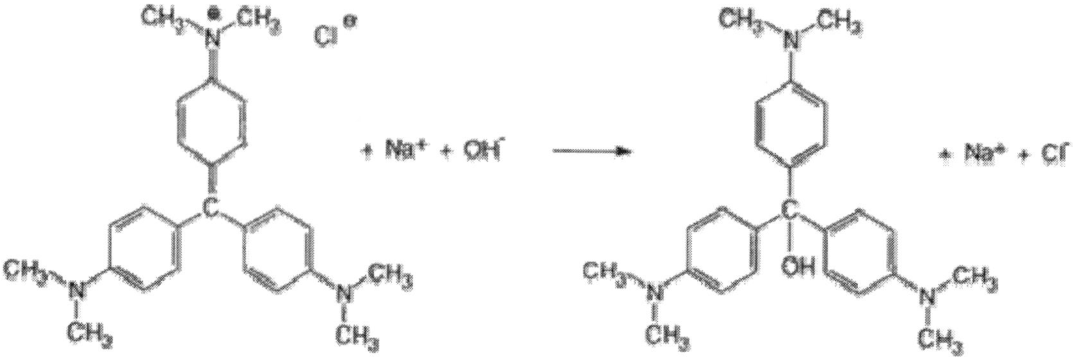

Figure 20.1. Molecular structure of crystal violet and the decolorization reaction with sodium hydroxide.

small whole numbers (usually, 0, 1 or 2). The objective of today's experiment will be to determine the numerical value of the rate constant, and the order of the chemical reaction with respect to both crystal violet (numerical of x) and hydroxide ion (numerical value of y).

INTEGRAL REACTION RATE METHOD

The rate constant and reaction orders can be determined by measuring the initial change in the dye concentration when the dye and hydroxide ion solutions are first mixed, or by measuring the dye concentration as the reaction proceeds to completion. Either method can be used; however, in this particular instance the latter method might be a little bit easier because the dye concentration can be determined by simply measuring the solution absorbance. Measuring the absorbance at several different times will give enough data to calculate k_{rate}, x and y.

Equation 20.1 is the differential form of the rate law expression. In this experiment we need to transform the rate law into its integral form. The transformation requires us to integrate Eqn. 20.1 over some definite time interval, lets say from an initial time when the chemicals are first mixed and both the dye and OH⁻ ion molar concentrations are known, to time t, where the dye concentration is determined by absorbance measurement. In order to perform the integration we need to make one additional stipulation. The OH⁻ ion must be held constant. Experimentally this is very easy to accomplish. We will simply prepare the solutions with a large excess of OH⁻

ions. The initial molar concentration will be much larger than the initial concentration of dye, *e.g.*, $[OH^-]_0 >> [Dye]_0$. The subscript "0" denotes the initial state at time equal to zero. Under this experimental condition, Eqn. 20.1 becomes

$$- d[Dye]/dt = k_{rate}' \, [Dye]^x \qquad (20.2)$$

where $k_{rate}' = k_{rate} [OH^-]^y$.

Equation 20.2 is integrated by assuming a numerical value for x. For a zero-order reaction (x = 0), integration gives

$$d[Dye]/dt = - k_{rate}' \, [Dye]^0 \qquad (20.3)$$

$$d[Dye] = - k_{rate}' \, dt \qquad (20.4)$$

$$\int_{[Dye]_0}^{[Dye]} d[Dye] = - \int_{t=0}^{t=t} k_{rate}' \, dt \qquad (20.5)$$

$$[Dye] - [Dye]_0 = - k_{rate}' \, t \qquad (20.6)$$

$$[Dye] = - k_{rate}' \, t + [Dye]_0 \qquad (20.7)$$

A plot of [Dye] versus time should be linear if the reaction of crystal violet with OH^- is zero-order. If the plot is not linear, one concludes that the reaction is not zero-order. Rather than calculating [Dye] for every absorbance measurement, Eqn. 20.7 will be expressed specifically in terms of absorbances:

$$A = - k_{rate}' \, \varepsilon_{Dye} \, b \, t + A_0 \qquad (20.8)$$

by substituting $[Dye] = A/(\varepsilon_{Dye} \, b)$. A linear relationship is still obtained, with the slope being negative and equal to $- k_{rate}' \, \varepsilon_{Dye} \, b$. Do not be concerned with understanding all of the details of the integration. The only information that is needed from the integration results is how to graph the experimental data for the various assumed reactions, and what does the slope equal, if a linear graph is observed. Equation 20.8 provides this information for a zero-order reaction.

The graphical form for a first-order reaction is obtained by setting x = 1

$$- d[Dye]/dt = k_{rate}' \, [Dye]^1 \qquad (20.9)$$

and then integrating the resulting equation

$$d[Dye]/[Dye] = - k_{rate}' \, dt \qquad (20.10)$$

$$\int_{[Dye]_0}^{[Dye]} d[Dye]/[Dye] = -\int_{t=0}^{t=t} k_{rate}{}' \, dt \qquad (20.11)$$

$$\ln\{[Dye]/[Dye]_0\} = -k_{rate}{}' \, t \qquad (20.12)$$

Expressed in terms of absorbances, the integral form of the first-order rate law equation for crystal violet gives

$$\ln A = -k_{rate}{}' \, t + \ln A_0 \qquad (20.13)$$

a linear relationship between the natural logarithm of measured values of absorbance and time. The slope of line is again negative, and the numeral value equals $-k_{rate}{}'$. The molar absorptivity of the dye, ε_{Dye}, cancels out in the final derived equation.

The final reaction order that will be considered is second-order, where $x = 2$ in Eqn. 20.1. After performing the required integration

$$-d[Dye]/dt = k_{rate}{}' \, [Dye]^2 \qquad (20.14)$$

$$d[Dye]/[Dye]^2 = -k_{rate}{}' \, dt \qquad (20.15)$$

$$\int_{[Dye]_0}^{[Dye]} d[Dye]/[Dye]^2 = -\int_{t=0}^{t=t} k_{rate}{}' \, dt \qquad (20.16)$$

a linear relation is

$$1/[Dye] = k_{rate}{}' \, t + 1/[Dye]_0 \qquad (20.17)$$

obtained between the reciprocal of the dye concentration and time. Expressed in terms of absorbances, A, Eqn. 20.17 becomes

$$1/A = [k_{rate}{}'/(\varepsilon_{Dye} \, b)] \, t + 1/A_0 \qquad (20.18)$$

If the reaction of crystal violet with hydroxide ion is second-order in dye, then a plot of reciprocal of the solution absorbance versus time should be linear with a positive slope. The numerical value of the slope equals $k_{rate}{}'/(\varepsilon_{Dye} \, b)$. Depending on which plot is linear, you may or may not need the molar absorptivity of crystal violet determined in Experiment 19. Assume that the path length for all of the measurements is 1 cm.

Once the first kinetic measurement is complete, the order of the chemical reaction with respect to crystal violet will be known. In other words, the numerical of "x" will be known from the graph that gave the best linear relationship. Use the correlation coefficient from the least squares analysis to determine which graph is the "most linear". The correlation coefficient nearest $r = 1$ (positive slope) or near $r = -1$ (negative slope) can be used to decide the reaction-order, should more than one of the graphs appear linear (see Figure 20.2 below). It is important that you make the absorption measurements for the specified time period. Even exponential curves appear linear when viewed over a very small distance or interval. The measurements need to be long enough for the nonlinearity to be observed. From the least squares analysis you will also have the numerical value of the slope, and the initial hydroxide ion concentration after mixing will be known. (The molar absorptivity coefficient and path length are also known, should the reaction turn out to be either zero-order or second-order in crystal violet.) You still need to calculate two values, the reaction-order with respect to OH⁻ ion (*e.g.*, value of "y") and k_{rate}. You have only one slope, however. The one slope equation contains two unknowns. You need to generate a second equation.

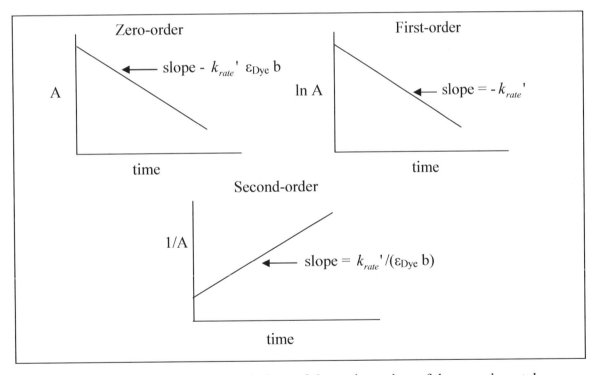

Figure 20.2 Graphical depictions of the various plots of the experimental absorbance versus time data for the determination of the reaction order with respect to crystal violet.

To get the second slope equation, you will need to perform the experiment a second time using a different hydroxide ion concentration. Once the experiment is done and analyzed, you will have sufficient experimental data for calculating y and k_{rate}. For illustrational purposes lets assume that in the first kinetic run, it was determined that a plog of ln A versus time was linear, and that the slope calculated through linear least squares analysis was 0.00015. The initial concentration of hydroxide after mixing is $[OH^-]_0$. Remember when chemicals are diluted/mixed, the new concentration is calculated by the formula:

$$\text{new concentration} = \text{old concentration} \times (\text{old volume/new volume}) \qquad (20.19)$$

The equation for the slope in the first with the numbers substituted in is:

$$k_{rate}' = 0.00015 = k_{rate} [0.005]^y \qquad (20.20)$$

assuming that the reaction is first-order in crystal violet. Slope equations for a zero-order reaction (see Eqn. 20.8) and for a second-order reaction (see Eqn. 20.18) will be different. Assume that the slope from the second kinetic run is 0.00029. The initial hydroxide ion concentration after mixing is $[OH^-] = 0.010$ for the second run. The equation for the slope for the second run is:

$$k_{rate}' = 0.00029 = k_{rate} [0.010]^y \qquad (20.21)$$

The value of y is calculated by dividing Eqn. 20.21 by Eqn. 20.20:

$$0.00029/0.00015 = (0.01/0.005)^y \qquad (20.22)$$

$$1.9333 = (2)^y \qquad (20.23)$$

Since the rate nearly doubled when the concentration of hydroxide ion was doubled, the reaction is first-order with respect to OH^-. Had the ratio of slopes been approximately equal, then $y = 0$. The reaction order would be second-order in OH^- had the ratio of slopes been 4.

Once the reaction order with respect to OH^- is known, the numerical value of k_{rate} rate is calculated simply as $k_{rate} = \text{slope}/[OH^-]^y$, using the values from either the first or second decolorization reaction. Every quantity on the right hand side of the equation is now known.

EXPERIMENTAL PROCEDURE FOR DECOLORIZATION REACTION

Set up the Pasco system to make absorbance measurements. Connect the Pasco Colorimeter to the PowerLink unit. The system should recognize that the colorimeter is connected when you first enter the system by clicking on the DataStudio icon and Create Experiment picture. Calibrate the colorimeter using deionized water, and change the default setting to display Green (565 nm) Absorbance. You will want to display the data graphically as a plot of absorbance versus time. (If you need to refer back to the instructions in Experiment 19 or in Appendix A.)

Once the colorimeter has been calibrated, you will need to transfer by buret 9.00 ml of 1.60×10^{-5} Molar crystal violet into a clean, dry 50-ml beaker. Next add 1.00 ml of 0.050 Molar sodium hydroxide to the crystal violet solution in the 50-ml beaker. Quickly mix by swirling (or by stirring with glass rod or coffee stirrer) without spilling. Fill the sample cell about 3/4ths full with solution, place the cell in the colorimeter sample holder, close the lid and **mouse click on the Start button**. Unlike in Experiment 13 it is not critical that the measurements be started simultaneously with mixing of the two solutions. A linear graph is linear no matter what y-value (or x-value) one starts at. The computation does not require the initial concentration of crystal violet, and in the case of the hydroxide ion the concentration is in such large excess that it remains constant for all practical purposes. Record the absorbance data for at least 30 minutes. While the experimental reaction is preceding record the molarity of the crystal violet and hydroxide ion on the laboratory Data Sheet in the section titled "First Crystal Violet Decolorization Reaction Data". Calculate the crystal violet and hydroxide ion concentrations after dilution using Eqn. 20.19.

When the 30 minutes have elapsed, **mouse click on the Stop button**. You are now ready to analyze the experimental data. Adjust the x- and y-scales so that the entire graph is visible on the computer screen. Refer to the instructions in the Appendix or in Experiment 8 for how to adjust the x- and y-scales. Sketch the graph of Absorbance versus time on the laboratory Data Sheet. To perform a linear least squares regression on the data **mouse click on the Fit ▼ button on the toolbar at the top of the graph screen. Scroll down to Linear Fit and mouse click**. The numerical value of the slope, intercept and correlation coefficient should appear on the computer screen. Record the information in the appropriate space on the Data Sheet.

Two more regressions still need to be performed. Clear the Linear fit curve from the computer screen by mouse clicking on Data ▼ button the top toolbar of the Graph screen. Scroll down to No Data and mouse click. If the original data disappears it can be retrieved by going back to the Data ▼ and marking Run 1 by mouse clicking to the left side. The data should now appear back on the graph. To perform the regression analysis for the first order rate law (*e.g.*, ln Absorbance versus time) **mouse click on the Calculator button** (first button to the right of Fit button on the graph top toolbar). In the definition box type y = . The rest of the box needs to be blank. Delete anything else that is in the definition box. Next **mouse click on Scientific ▼ box** and scroll down the list of options until the ln(x) line is highlighted. Now you need to tell the system where to find the x-values that it is to take the natural logarithm of. **Click on the ▼ in the Variables box**, scroll down the list of options, and **mouse click when Data Measurement is highlighted**. A new screen should open up. You will want to **mark the run under the Green (565 nm) Absorbance**. At this time there should be only one run present, however, after the second decolorization run, you will have two sets of experimental data. Scroll down to mark the set of data that you want to use in the calculation. **Mouse click OK and then Yes** on the next screen. This should return you back to the Calculator window. **Mouse click Accept** and **close the calculator window**. The calculated experimental data should now be graphed on the computer screen. To perform the regression analysis, go up to the **Fit ▼ on the top of the Graph toolbar, scroll down to Linear Fit and mouse click**. A box of statistical information should appear on the computer screen giving the slope, intercept and correlation coefficient. There should be an arrow pointing from the box to the curve that the information pertains to. Record this information on the laboratory Data Sheet.

To get back to the original Absorbance versus time data, close the graph screen by **mouse clicking on the X box in the upper right-hand corner of the graph screen**. (Be careful not to close the entire system.) To the left of the computer screen, about midway down the left-hand side, there will be a window labeled Display. **Double left-hand mouse click on the line corresponding to Graph**. A window should open up on the right. **Scroll down to Green (565 nm) Absorbance line and mouse click**. To perform the second-order regression analysis, **click on the Calculator button** on the top toolbar of the graph screen while the Absorbance versus time data is being displayed. In the Definitions box type y = 1/x. (There is no pull down option for this calculation so you have to type the entire calculation. Delete anything else from the Definitions box.) **Next mouse click on the ▼ in the Variables box, scroll down to Data Measurement and mouse click again. Highlight Green (565 nm) Absorbance line and mouse click. Click Yes** when the screen comes up informing you that the calculation is being performed on only one set of data. At this point in time you should now be back to the Calculator screen. **Mouse click on Accept**, and **close the calculator box**. The graph of the experimental data as 1/Absorbance versus time should now be displayed, along with the Absorbance versus time data. To perform a linear least squares fit on the

1/Absorbance versus time data, go to **the Fit ▼ on the top toolbar of the graph window, scroll down to linear fit and mouse click.** The statistical information should appear in a box. Record the information on your laboratory Data Sheet.

All of the pertinent experimental data should now be recorded on the laboratory Data Sheet for the first crystal violet decolorization measurement. It is now time to perform the experiment measurement again, this time using a different concentration of sodium hydroxide. The second set of experimental data can be recorded right over the first set. Close all of the graph screens, and **double left-hand mouse click on the Graph line in the Display window** on the left-hand side of the computer screen. **Scroll down to Green (565 nm) Absorbance and mouse click** when this line is highlighted. The absorbance data for the first crystal violet decolorization experiment should be graphically displayed. Transfer by buret 9.00 ml of 1.60×10^{-5} Molar crystal violet into a clean, dry 50-ml beaker. Next add 1.00 ml of 0.10 Molar sodium hydroxide to the crystal violet solution in the 50-ml beaker. **(Make sure that you use the 0.10 Molar sodium hydroxide for the second decolorization reaction.)** Quickly mix by swirling (or by stirring with glass rod or coffee stirrer) without spilling. Fill the sample cell about 3/4ths full with solution, place the cell in the colorimeter sample holder, close the lid and **mouse click on the Start button.** While the experimental reaction is preceding record the molarity of the crystal violet and hydroxide ion the laboratory Data Sheet in the section titled "First Crystal Violet Decolorization Reaction Data". Calculate the crystal violet and hydroxide ion concentrations after dilution using Eqn. 20.19. When the 30 minutes have elapsed, **mouse click on the Stop button.** Perform the Absorbance versus time, Ln Absorbance versus time, and 1/Absorbance versus time linear regression analysis on the experimental data for the second decolorization reaction. You really should check all graphs to make sure that the reaction mechanism has not changed. For some experimental reactions, the reaction mechanism may change at higher/lower reactant concentrations. That should not be the case in today's experiment.

After all six linear regressions have been performed you need to decide what the reaction order is with respect to crystal violet. Which way of graphing the experimental data gave the "best" straight line? Visually look at the graphs and the corresponding correlation coefficients that you recorded on the laboratory Data Sheet. Once the reaction order is known for crystal violet, the reaction order with respect to hydroxide ion is determined using the slopes from the lines for the first and second crystal violet decolorization reactions. The slope equation for a zero-order reaction is slope = $- k_{rate}'$ ε_{Dye} b, for a first-order reaction is slope = $-k_{rate}'$, and for a second-order reaction is slope = $k_{rate}'/(\varepsilon_{Dye}$ b). You will need to use the molar absorptivity coefficient, ε, from last week's experiment if the reaction order turns out to be either zero-order or second-order in crystal violet. The molar absorptivity coefficient should be listed on the Data Sheet as item number 1. Assume that the path length is $b = 1$ cm for all calculations. Record the numerical values of k_{rate}' for both the first and second decolorization reactions on lines 8 and 15 of the Data Sheet. The reaction order with respect to OH⁻ is calculated by substituting the two k_{rate}' values and [OH⁻] concentrations into $k_{rate}' = k_{rate} [OH^-]^y$. (Follow the example calculation given by Eqns. 20.20 – 20.23.)

Once the reaction order with respect to OH⁻ is known, the numerical value of k_{rate} rate is calculated simply as $k_{rate} = slope/[OH^-]^y$, using the values from either the first or second decolorization reactions. Every quantity on the right had side of the above equation is now known.

There is a published literature value of $k_{rate} = 0.194$ dm³/(mol sec) for the rate constant for the decolorization of crystal violet by hydroxide ion (see S. F. Beach, J. D. Hepworth, D. Mason and E. A. Swarbrick, *Dyes and Pigments*, **42**, 71-77 (1999). How does your measured rate constant compare to the literature value?

DATA SHEET – EXPERIMENT 20

crystal violet

Name: _____

First Crystal Violet Decolorization Reaction Data

1. Molar absorptivity coefficient of crystal violet, ε,
   from Experiment 19, liters/(mole cm):   _____

2. Concentration of crystal violet before dilution, M: _____

3. Concentration of crystal violet after dilution, M: _____

4. Concentration of sodium hydroxide before dilution, M: _____

5. Concentration of sodium hydroxide after dilution, M: _____

6. Sketch of graphical results with linear least squares regression information:

   Absorbance *versus* time   Ln Absorbance *versus* time

   slope = _____   slope = _____

   corr. coeff. = _____   corr. coeff. = _____

   1/Absorbance *versus* time

   slope = _____

   corr. coeff. = _____

7. Order of reaction with respective to crystal violet: _____

8. Calculated value of k_{rate}' from slope: _____

Second Crystal Violet Decolorization Reaction Data

9. Concentration of crystal violet before dilution, M: _____

10. Concentration of crystal violet after dilution, M: _____

11. Concentration of sodium hydroxide before dilution, M: _____

12. Concentration of sodium hydroxide after dilution, M: _____

13. Sketch of graphical results with linear least squares regression information:

Absorbance *versus* time

slope = _____

corr. coeff. = _____

Ln Absorbance *versus* time

slope = _____

corr. coeff. = _____

1/Absorbance *versus* time

slope = _____

corr. coeff. = _____

14. Order of reaction with respective to crystal violet: _____

15. Calculated value of k_{rate}' from slope: _____

16. Order of reaction with respect to hydroxide ions: _____

Rate Constant Calculations

17. Calculated value of k_{rate} from first decolorization reaction: _____

18. Calculated value of k_{rate} from second decolorization reaction: _____

19. Mean value of rate constant: _____

20. Units for the rate constant: _____

21. Does the order of the reaction with respect to both crystal violet and hydroxide ion suggest a simple bi-molecular reaction mechanism, or a more complicated reaction mechanism involving more than a single step. Explain your answer in essay style format.

EXPERIMENT 21: SPECTROMETRIC DETERMINATION OF AN EQUILIBRIUM CONSTANT FOR COMPLEX FORMATION

INTRODUCTION

Chemical equilibrium results from the fact that many chemical reactions can proceed simultaneously in opposite directions. Take for example the general chemical reaction between chemicals A and B to form products C and D:

Chemical A + Chemical B   <------>   Chemical C + Chemical D

with the bi-directional arrow, " <------> ", indicating that the reaction occurs in both the forward and reverse direction. If C and D are the first and only products formed by the reaction of chemicals A and B, then the rate of reaction depends on the molar concentrations of A and B. As the reaction proceeds from left to right, the reactant concentrations decrease, thus causing a simultaneous decrease in the rate of reaction. Initially, chemicals C and D were not present, so there was no opposite or reverse reaction. However, once chemicals C and D are formed, and their concentrations increase as the result of the forward reaction, the rate of the reverse reaction increases with increasing C and D concentration. If the forward rate of reaction is decreasing from its initial value, while the reverse reaction rate is increasing, there will eventually be a point in time when the two reaction rates are equal. The system is now at chemical equilibrium. Once chemical equilibrium is established, there is no net change in the concentrations of any of the species.

The objective of today's laboratory experiment is to measure the equilibrium constant for the red complex formed from the reaction of iron(III) and thiocyanate ions:

$$Fe^{3+}_{(aq)} + SCN^{-}_{(aq)} \quad \text{<------>} \quad Fe(SCN)^{2+}_{(aq)}$$

At very low thiocyanate to iron(III) concentration ratios, only the 1:1 complex is formed. As the relative concentration of SCN^- increases, additional complexes are formed. The 1:2 iron(III)-thiocyanate complex, $Fe(SCN)_2^+$, starts to form when the SCN^- to Fe^{3+} ratio is slightly above 4. At very large SCN^- concentrations, the iron-thiocyanate complex that is in solution is $Fe(SCN)_6^{3-}$. This complex is also red in color. To prevent the formation of higher-order complexes, you will be using solutions in which $[Fe^{3+}] \geq [SCN^-]$.

From a mathematical standpoint, the equilibrium condition for the formation of the $Fe(SCN)^{2+}$ complex is described the equilibrium constant

$$K_{eq} = \frac{[Fe(SCN)^{2+}]}{[Fe^{3+}][SCN^-]} \qquad (21.1)$$

which is a ratio of the molar concentration of the complex divided by the product of the molar concentrations of the two reactants. The concentrations in Eqn. 21.1 refer to the values after equilibrium has been established, and not the initial stoichiometric concentrations used in preparing the solution. To calculate the equilibrium constant one needs to substitute numerical values for the equilibrium concentrations of $Fe(SCN)^{2+}$, Fe^{3+} and SCN^{-} ions.

In principle the concentration of the red $Fe(SCN)^{2+}$ complex can be determined by measuring the amount of light that passes through the solution. The maximum absorption of $Fe(SCN)^{2+}$ is around 447 nm (blue light), and neither Fe^{3+} nor SCN^{-} absorb appreciable radiation in this spectral region. (Higher-order complexes like $Fe(SCN)_2^{+}$, ..., $Fe(SCN)_6^{3-}$ would absorb, that is one of the reason for keeping $[Fe^{3+}] \geq [SCN^{-}]$.) From Experiment 19 one learned that the measured absorbance, A, is

$$A = constant \times C = \varepsilon\, b \times C \qquad (21.2)$$

directly proportional to the molar concentration of the radiation-absorbing species (chromophore). If the Beer's law proportionality constant can be determined for the $Fe(SCN)^{2+}$ complex, then this value can be used to determine the molar concentrations of $Fe(SCN)^{2+}$ in the equilibrium solutions that you will study.

Determining the proportionality constant is not as simple as it was in the case of the dye solutions studied in Experiment 19. If one were to dissolve a known amount of $Fe(SCN)^{2+}$ in solution, part of the complex would back dissociate into Fe^{3+} and SCN^{-} ions, and the molar concentration of $Fe(SCN)^{2+}$ of would no longer by known. To calculate the equilibrium value of $[Fe(SCN)^{2+}]$ needed in Eqn. 21.2 for the Beer's law proportionality constant determination, one must know the value of K_{eq}. To calculate the K_{eq} from spectroscopic measurements, one would need to know the Beer's law proportionality constant. Initially it may appear that there is no way around this dilemma, fortunately there is, otherwise it would not be possible to determine equilibrium constants for such cation-anion complexes.

Lets ask the following question, is there any set of experimental conditions that would suppress the back dissociation of the $Fe(SCN)^{2+}$ complex? Yes, one could apply Le Chatelier's Principle. A large excess of one of the reactants should force the equilibrium to the right-hand (products) side. A large excess of SCN^{-} cannot be tolerated as this would lead to the formation of additional Fe-SCN complexes, so the solutions will be prepared with a large excess of Fe^{3+} for the Beer's law proportionality constant determination.

To illustrate the computational method, we will consider the first solution that you will make that contains both Fe^{3+} and SCN^{-} ions. The first solution is prepared by mixing 2.5 ml of 0.200 Molar $Fe(NO_3)_3$ with 0.1 ml of 0.002 Molar NaSCN and 7.4 ml of 0.10

Molar HNO_3. The nitric acid is added to obtain an acidic solution, otherwise Fe^{3+} ions would precipitate from solution as solid $Fe(OH)_3$. A three-line table is set up

	$Fe^{3+}_{(aq)}$	+	$SCN^-_{(aq)}$	------>	$Fe(SCN)^{2+}_{(aq)}$
Initial:	(2.5/10)(0.2)		(0.1/10)(0.002)		0
Reacts:	- (0.1/10)(0.002)		- (0.1/10)(0.002)		(0.1/10)(0.002)
At equilibrium:	$\approx$ (2.5/10)(0.2)		0		(0.1/10)(0.002)

assuming that the Fe^{3+} concentration is sufficiently large to force the reaction to go completely to the right-hand side. The bi-directional arrow (<------>) has been replaced with a unidirectional arrow (------>) in order to emphasize the fact that there is a very large excess of Fe^{3+} in solution, forcing the reaction to the right-hand side. Note, the equilibrium concentrations that appear in the first row of the table have been corrected for dilution

$$\text{new concentration} = \text{old concentration} \times (\text{old volume/new volume}) \qquad (21.3)$$

Once the absorbance of this solution is measured, you will have one experimental (concentration, absorbance) ordered data point pair for determining the Beer's law proportionality constant.

Several additional data points will be obtained by preparing solutions that contain 0.2, 0.4, 0.6, 0.8 and 1.0 ml of 0.002 Molar NaSCN. Each solution will still contain 2.5 ml of 0.200 Molar $Fe(NO_3)_3$, and the volume of 0.10 Molar HNO_3 will be adjusted so that the total volume of each solution prepared remains at 10.0 ml. Linear least squares analyses of the solution absorbance (y-axis) versus $[Fe(SCN)^{2+}]$ (x-axis) should give a straight line, with the slope of the line being equal to the Beer's law proportionality constant, e.g., slope = constant = ε b (see Eqn. 21.2).

Beer's law will then provide the means to measure the equilibrium molar concentration of the $Fe(SCN)^{2+}$ complex in the second series of solutions that you will prepare. (The proportionality constant will now be known from the first series of solutions.) For the equilibrium constant determination one must have enough Fe^{3+}, SCN^- and $Fe(SCN)^{2+}$ in the solution so that their concentrations can be measured and/or calculated with a large level of certainty. Solutions containing an extremely large excess of Fe^{3+} cannot be tolerated because this would give a numerical value of $[SCN^-] \approx 0$ at equilibrium. There would be too much uncertainty in the value of $[SCN^-]$ at equilibrium to permit a "good" calculation of K_{eq}. For equilibrium constant determinations one generally uses reactant concentrations close to the stoichiometric combining ratio, which in our case is 1:1. The $[Fe^{3+}]$ to $[SCN^-]$ ratios that will be used in the K_{eq} determination range between $[Fe^{3+}]/[SCN^-] = 5$ to $[Fe^{3+}]/[SCN^-] = 1$.

The iron(III) concentration will no longer be in large excess in the second series of solutions, so the reaction must be treated mathematically as an equilibrium. The first solution in the second series that will contain complex will be prepared by missing 5.0 ml of 0.00200 Molar $Fe(NO_3)_3$ with 1.0 ml of 0.00200 Molar NaSCN and 4.0 ml of 0.10 Molar HNO_3. In setting up the three-line table

	$Fe^{3+}_{(aq)}$	+	$SCN^-_{(aq)}$	<------>	$Fe(SCN)^{2+}_{(aq)}$
Initial:	(5.0/10)(0.002)		(1.0/10)(0.002)		0
Reacts:	-x		- x		x
At equilibrium:	0.001 - x		0.0002 - x		x

The first row is again the initial molar concentrations after mixing. The amounts of Fe^{3+} and SCN^- that react to form complex are not known; hence "x" appears in all three columns of the second row. The third row in the table corresponds to the concentrations that are in the solution at equilibrium. It is these concentrations that must be substituted into Eqn. 21.1

$$K_{eq} = \frac{x}{[(5/10)(0.002) - x][(1/10)(0.002) - x]} \qquad (21.4)$$

to calculate the equilibrium constant. The concentration of the complex in solution is determined by measuring the absorbance of light, and then dividing this number by the Beer's law proportionality constant from the first series of solutions, e.g., $[Fe(SCN)^{2+}]$ = absorbance of solution/εb.

A numerical value has now been calculated for K_{eq}, but is this value really an equilibrium constant? What must one experimentally show in order to "prove" that the concentration ratio defined by Eqn. 21.1 is indeed an equilibrium constant? One must verify that the same numerical value is obtained with other Fe^{3+} and SCN^- concentrations. Four additional solutions will be prepared by mixing 5.0 ml of 0.0020 Molar $Fe(NO_3)_3$ with 2.0, 3.0, 4.0 and 5.0 ml of 0.0020 Molar NaSCN for this verification. The volume of 0.10 Molar HNO_3 will be adjusted so that the total volume of each solution is 10 ml.

In preparation for this laboratory you can calculate and record the equilibrium molar concentrations of $Fe(SCN)^{2+}$ (item # 2 on the laboratory Data Sheet) for the Beer's Law proportionality constant determination, and the initial molar concentrations of Fe^{3+} and SCN^- (item #s 7 and 8 on the Data Sheet) prior to attending laboratory. The concentrations are based Eqn. 21.3. Table 21.1 (item #2) and Table 21.2 (item #s 6 and

7) describe how the various solutions are prepared. By doing the calculations ahead of time, you should not have to spend as much time physically in the laboratory.

EXPERIMENTAL DETERMINATION OF BEER'S LAW PROPORTIONALITY CONSTANT

Using a pipet or buret, add the volumes of 0.200 Molar $Fe(NO_3)_3$ and 0.0020 Molar NaSCN stock solutions indicated in Table 21.1 into seven clean, dry 50-ml or 100-ml beakers. (Both stock solutions were prepared in 0.10 Molar HNO_3 in order to maintain an acidic mixture, which will prevent $Fe(OH)_3$ from precipitating out of solution.) Also pipet the indicated volume of 0.10 Molar HNO_3 stock solution into the respective beakers to bring the total volume up to 10 ml. Stir the mixtures thoroughly. Calculate the molar concentration of SCN^- in each solution after mixing using the dilution equation, Eqn. 21.3. The concentration of SCN^- after mixing is the equilibrium value of $[Fe(SCN)^{2+}]$ that will be used in the linear least squares regression analysis for the Beer's law proportionality constant determination. Record the values of $[Fe(SCN)^{2+}]$ on the laboratory Data Sheet under the section titled "Beer's Law Proportionality Constant Determination".

Once the seven solutions listed in Table 21.1 have been prepared, you are now ready to begin making the absorbance measurements. Enter the Pasco system by double left-hand mouse clicking on the DataStudio icon that appears on the computer screen. You will need to calibrate the Pasco Colorimeter with the solution in Beaker # 1. A solution of iron(III) is yellowish-orange in color, and one needs to eliminate any absorbance due to iron(III) from the measured absorbance. You want the entire measured absorbance to be due to the Fe-SCN complex. Put the solution in Beaker 1 into the sample cuvette, place the cuvette into the Colorimeter sample holder, close the lid, and press the green oval calibrate button. Once the green light next to the green oval goes out the Colorimeter is calibrated. Go into the Setup menu, and change the display output to read the Blue (468 nm) Absorbance. Refer back to the instructions in Experiment 19 if you need to. Once the system is setup measure the absorbances of the solutions in

Table 21.1. Preparation of Solutions for the Beer's Law Proportionality Constant Determination

Beaker Number	0.200 M Fe(NO₃)₃ in 0.10 M HNO₃	0.00200 M NaSCN in 0.10 M HNO₃	0.10 Molar HNO₃
1**	2.5 ml	0.0 ml	7.5 ml
2	2.5 ml	0.1 ml	7.4 ml
3	2.5 ml	0.2 ml	7.3 ml
4	2.5 ml	0.4 ml	7.1 ml
5	2.5 ml	0.6 ml	6.9 ml
6	2.5 ml	0.8 ml	6.7 ml
7	2.5 ml	1.0 ml	6.5 ml

** This solution will be used to calibrate the colorimeter.

Beakers 2 – 7, and record the experimental values on the laboratory Data Sheet in the section titled "Beer's Law Proportionality Constant Determination." The proportionality constant, ε, is obtained through a linear least squares analysis of the absorbance data for the solutions in beakers 2-7 (y-axis) plotted against the molar concentration of the $FeSCN^{2+}$ complex, *e.g.*, A versus $[FeSCN^{2+}]$. The path length is assumed to be one, b = 1 cm. There is an published experimental value of $\varepsilon = 1.11 \times 10^5$ liter/(mole cm) in the literature for the $FeSCN^{2+}$ complex (K. Khan, H. Masood and T. A. Khan, *Pakistan Council of Scientific and Industrial Research*, **41**, 217-220 (1998))

EXPERIMENTAL DETERMINATION OF EQUILIBRIUM CONSTANT FOR Fe(SCN)²⁺ COMPLEX

Using a pipet or buret, add the volumes of 0.0020 Molar Fe(NO₃)₃ and 0.0020 Molar NaSCN stock solutions indicated in Table 21.2 into seven clean, dry 50-ml or 100-ml beakers. (Both stock solutions were prepared in 0.10 Molar HNO₃ in order to maintain an acidic mixture, which will prevent Fe(OH)₃ from precipitating out of solution.) Also pipet the indicated volume of 0.10 Molar HNO₃ stock solution into the respective beakers to bring the total volume up to 10 ml. Stir the mixtures thoroughly. Calculate the "initial" molar concentrations of Fe^{3+} and SCN^- in each solution after mixing using the dilution equation, Eqn. 21.3. Enter these values under item # 6 and # 7 on the laboratory Data Sheet in the section titled "Spectroscopic Determination of Equilibrium Constant for $Fe(SCN)^{2+}$".

Table 21.2. Preparation of Solutions for the Equilibrium Constant Determination

Beaker Number	0.0020 M $Fe(NO_3)_3$ in 0.10 M HNO_3	0.00200 M NaSCN in 0.10 M HNO_3	0.10 Molar HNO_3
8**	5.0 ml	0.0 ml	5.0 ml
9	5.0 ml	1.0 ml	4.0 ml
10	5.0 ml	2.0 ml	3.0 ml
11	5.0 ml	3.0 ml	2.0 ml
12	5.0 ml	4.0 ml	1.0 ml
13	5.0 ml	5.0 ml	0.0 ml

** This solution will be used to calibrate the colorimeter.

Once all of the solutions are thoroughly mixed, you are ready to start making the absorbance measurements. The Pasco Colorimeter does have to be recalibrated due to the fact that the solutions contain a significantly less Fe^{3+}. Recalibrate the colorimeter by pouring the solution in beaker number 8 into the sample cell. Place the sample cell into the holder, close the holder lid, and then press the green oval calibrate button. After the green light goes out on the colorimeter, the system should be calibrated for the second set of measurements. Go into the Setup menu to change the display output to read the Blue (468 nm) Absorbance. Measure the absorbances of solutions 9 – 13, and record the experimental data in the appropriate place on the Data Sheet. Calculate the equilibrium concentration of $[Fe(SCN)^{2+}]$ in the five solutions by dividing the measured absorbance by the numerical value of the molar absorptivity coefficient that you determined from the first series of absorbance measurements. The equilibrium molar concentrations of Fe^{3+} and SCN^- are obtained by $[Fe(SCN)^{2+}]$ from the initial molar concentrations of the ions after mixing (item #s 6 and 7). The equilibrium molar concentrations of the three ions are then substituted into Eqn. 21.1. You should have a calculated equilibrium constant for each mixture (solutions 9 – 13). The calculated values should be approximately equal, at least to within experimental uncertainty.

Calculate the average value of the equilibrium constant and the standard deviation, and give your average value of K_{eq} to the TA who will then compile the values for the entire class. The data for the entire class will be posted on the bulletin board outside the laboratory room sometime during the next day. Go to the bulletin board outside the laboratory room and copy down the class results on your laboratory Data Sheet in the space provided. Calculate the class mean (average) and standard deviation for the equilibrium constant determination.

DATA SHEET – EXPERIMENT 21

Pasco colorimeter

Name: _____

Beer's Law Proportionality Constant Determination

1. Measured solution absorbance:

_____   _____   _____   _____   _____   _____

Beaker 2   Beaker 3   Beaker 4   Beaker 5   Beaker 6   Beaker 7

2. Equilibrium concentration of $Fe(SCN)^{2+}$, M:

_____   _____   _____   _____   _____   _____

Beaker 2   Beaker 3   Beaker 4   Beaker 5   Beaker 6   Beaker 7

3. Beer's law proportionality constant, εb, from linear regression: _____

4. Correlation coefficient from linear least squares regression: _____

Spectroscopic Determination of Equilibrium Constant for $Fe(SCN)^{2+}$

5. Measured solution absorbance:

_____   _____   _____   _____   _____

Beaker 9   Beaker 10   Beaker 11   Beaker 12   Beaker 13

6. Initial concentration of Fe^{3+}, M:

_____   _____   _____   _____   _____

Beaker 9   Beaker 10   Beaker 11   Beaker 12   Beaker 13

7. Initial concentration of SCN^-, M:

_____   _____   _____   _____   _____

Beaker 9   Beaker 10   Beaker 11   Beaker 12   Beaker 13

8. Equilibrium concentration of $Fe(SCN)^{2+}$ from absorbance measurements, M:

    _____   _____   _____   _____   _____

    Beaker 9   Beaker 10   Beaker 11   Beaker 12   Beaker 13

9. Equilibrium concentration of Fe^{3+}, M:

    _____   _____   _____   _____   _____

    Beaker 9   Beaker 10   Beaker 11   Beaker 12   Beaker 13

10. Equilibrium concentration of SCN^-, M:

    _____   _____   _____   _____   _____

    Beaker 9   Beaker 10   Beaker 11   Beaker 12   Beaker 13

11. Calculated values of equilibrium constant, K_{eq}, for $Fe(SCN)^{2+}$ formation:

    _____   _____   _____   _____   _____

    Beaker 9   Beaker 10   Beaker 11   Beaker 12   Beaker 13

12. Average value of equilibrium constant: _____

13. Standard deviation of equilibrium constant: _____

14. Literature value of equilibrium constant: 1.05×10^3

15. Discuss how your experimental value compares to the literature value.

Class Data – Equilibrium Constant for Fe(SCN)$^{2+}$ Complex:

_____	_____	_____	_____
_____	_____	_____	_____
_____	_____	_____	_____
_____	_____	_____	_____
_____	_____	_____	_____

16. Class Mean Equilibrium Constant: _____

17. Class Standard Deviation: _____

EXPERIMENT 22: MOLAR SOLUBILITY AND DETERMINATION OF SOLUBILITY PRODUCT

INTRODUCTION

Solubility is defined as the maximum amount of a solute that will dissolve in a given quantity of solvent at a particular temperature and pressure to form a saturated solution. As you observed in Experiment 2 (see Table 2.1) some pairs of liquids such as water and acetone will mix in all proportions (completely miscible), whereas other pairs of liquids like hexane and water do not dissolve in one another (immiscible). The extent to which one substance dissolves in another depends on the nature of both the solute and the solvent. Strong attractive forces between the solute and solvent favor dissolution. Substances tend to be immiscible whenever the attractions between the solute and surrounding solvent molecules are much weaker than the solute-solute and solvent-solvent interactions. Observations have shown that two compounds with similar inter-molecular attractive forces tend to be soluble in one another. Nonpolar solutes are more likely to dissolve in nonpolar solvents; polar solutes and ionic compounds are more likely to be soluble in polar solvents.

Ionic compounds dissolve in water to form hydrated ions in solution according to the following equilibria

$$(Cation_xAnion_y)_{(solid)} \quad \longleftrightarrow \quad x\,Cation^{y+}_{(aq)} + y\,Anion^{x-}_{(aq)}$$

As with any equilibrium, the extent to which this dissolution reaction occurs is expressed by the magnitude of the equilibrium constant

$$K_{eq} = K_{sp} = [Cation^{y+}]^x\,[Anion^{x-}]^y \qquad (22.1)$$

Solids and the solvent do not appear in the equilibrium constant expression. The equilibrium constant is called the solubility product, K_{sp}. The numerical value of K_{sp} can be calculated by measuring the molar concentrations of the cation and anion in the saturated solution. In more advanced level chemistry courses you will learn that the solubility product equals the product of the *thermodynamic concentrations* (called activities) of the cation and anion involved in the equilibrium, each concentration raised to the power of its respective coefficient in the balanced chemical equilibrium. Molarities and thermodynamic activities are taken as equivalent quantities in introductory general chemistry courses.

In today's laboratory experiment the solubility product for the dissociation of calcium hydroxide

$$Ca(OH)_{2(solid)} \quad \longleftrightarrow \quad Ca^{2+}_{(aq)} + 2\,OH^-_{(aq)}$$
$$K_{sp} = [Ca^{2+}]\,[OH^-]^2 \qquad (22.2)$$

will be determined by measuring the concentration of the hydroxide ion in the saturated solution. The concentration of OH⁻ can be determined by measuring the pH of the saturated solution with a pH probe/electrode

$$pH = -\log\ [H^+] \tag{22.3}$$

$$[H^+] = 10^{-pH} \tag{22.4}$$

$$[OH^-] = 10^{-14}/[H^+] \tag{22.5}$$

or by titrating a known volume of the saturated solution with a strong acid, such as hydrochloric acid:

$$OH^-_{(aq)} + HCl_{(aq)} \longrightarrow H_2O + Cl^-_{(aq)}$$

$$\text{moles of OH}^- = \text{moles of HCl} \tag{22.6}$$

$$\text{molarity of OH}^- \times \text{volume of OH}^- = \text{molarity of HCl} \times \text{volume of HCl} \tag{22.7}$$

For notational simplicity the short-hand notation of H^+ (rather than H_3O^+) is used. Since the saturated solution was prepared by adding only solid $Ca(OH)_2$, there would be one Ca^{2+} ion for every 2 OH⁻ ions, *e.g.*, $[Ca^{2+}] = 0.5\ [OH^-]$. The amount of OH⁻ coming from the dissociation of water will be negligible. Both methods will be used, so that the measured values can be compared.

Titrimetric methods generally give better results for this type of experimental determination. Errors in concentrations are generally less than ± 1% if the titration is carefully performed. Errors in pH measurements are generally much larger, particularly in the case of highly alkaline solutions. The reason for the much larger error is related to how the glass pH electrode works and to problems associated with finding pH calibration standards for pHs > 10.

The preceding discussion is a simplified treatment of the solubility phenomena. Many sparingly soluble salts do not completely dissociate into their constituent ions, but interact in solution with one another to form soluble aggregates, called complexes, or react with water to form new species. For example, in a saturated solution of calcium sulfate, $CaSO_4$, a significant fraction of the dissolved calcium sulfate exists as molecular $CaSO_{4(aq)}$. In saturated calcium sulfate solutions it takes two equilibrium (and two equilibrium constant expressions) to describe adequately the system:

$$CaSO_{4(solid)} \longleftrightarrow Ca^{2+}_{(aq)} + SO_4^{2-}_{(aq)}$$

$$K_{sp} = [Ca^{2+}][SO_4^{2-}] \tag{22.8}$$

$$CaSO_{4(solid)} \longleftrightarrow CaSO_{4(aq)}$$

$$K_{intrinsic} = [CaSO_{4(aq)}] \tag{22.9}$$

The measured solubility is the sum of the amount of solid that ionizes plus the intrinsic solubility (*e.g.*, $CaSO_{4(aq)}$ in solution).

By comparing the experimental solubility product that you determine today for $Ca(OH)_2$ based on the HCl titration data to the more carefully determined literature value of $K_{sp} = 6.5 \times 10^{-6}$ (value is for 25 °C) you should be able to ascertain whether or not the intrinsic solubility and/or complex formation (such as the formation of $CaOH^+$) is important. Both effects would cause the measured K_{sp} to be too large. Researchers would have removed the intrinsic solubility contribution, as well as any complex formation, from the literature K_{sp} value using a more complete thermodynamic treatment.

ESTIMATION OF THE SOLUBILITY PRODUCT OF CALCIUM HYDROXIDE USING *pHydrion* PAPER

Pour approximately 80 ml of saturated calcium hydroxide solution into a 100 ml beaker. This should be enough calcium hydroxide for you to complete the entire experiment. Be careful not to disturb the solid residue at the bottom the reagent bottle. You want to analyze only the calcium hydroxide that actually dissolved, not any suspended solid particles. Dip the tip of small strip of *pHydrion* paper (pH 1 to 13 range) inot the solution. The tip is coated with pigment dyes that turn various colors at the different pHs. Estimate the pH of the saturated, as best as you can, by comparing the tip of the strip to the color chart that is provided for you to use. Record the estimated pH on your laboratory Data Sheet in the section labeled "Solubility Product Estimation Using pH Paper". Now estimate the solubility product of calcium hydroxide, using your estimated pH and Eqns. 22.2 – 22.5. Remember $[Ca^{2+}] = 0.5\ [OH^-]$. The term *estimate* is deliberately used for this portion of today's laboratory exercise as it is very unlikely that the calculated value of K_{sp} based on the color of the *pHydrion* paper will be very close to the accepted value of $K_{sp} = 6.5 \times 10^{-6}$. The estimated value though is better than the experimental value you had prior to dipping the pH paper in the saturated calcium hydroxide solution, which was no idea of what the value should be, except for the literature value that was given to you for comparison purposes.

DETERMINATION OF THE SOLUBILITY PRODUCT OF CALCIUM HYDROXIDE USING A GLASS pH ELECTRODE

For this part of the laboratory experiment you will need to dilute the saturated calcium hydroxide solution with deionized water. Recall that the uncertainty in the pH measured with a glass membrane pH electrode increases at pH > 9. Transfer 5.0 ml of the saturated calcium hydroxide solution (by pipet) into a 250-ml beaker. Now add 95 ml of deionized water (graduated cylinder). [If available a 100-ml volumetric flask would be better for making this dilution.] Stir thoroughly to mix. Now measure the pH of the solution with the Pasco pH electrode. Record the pH on the laboratory sheet in the appropriate section of the Data Sheet. Calculate $[H_3O^+]$ in the solution by substituting the measured value of the pH into Eqn. 22.3, which is then inserted into Eqn. 22.5 to get $[OH^-]$ in the diluted solution. You will now need to multiply the value by 20 in order to undo the dilution that was made when 5 ml of saturated calcium hydroxide was mixed with 95 ml of deionized water. After multiplying by 20, insert the corrected value of $[OH^-]$ into Eqn. 22.2 to calculate the solubility product. How does your calculated value of K_{sp} compare to the accepted value of $K_{sp} = 6.5 \times 10^{-6}$? [Item to think about – Why does one first calculate the hydroxide ion concentration and then multiply by 20 to undue the dilution that was made? Why not multiply the hydronium ion concentration by 20 to undo the dilution, and then calculate the hydroxide ion concentration using Eqn. 22.5?]

DETERMINATION OF SOLUBILITY PRODUCT BY SODIUM HYDROXIDE TITRATION

The last method that will be used to determine the solubility product of calcium hydroxide will involve titration of an aliquot of the saturated calcium hydroxide solution with sodium hydroxide. Rinse and fill the buret with 0.02 Molar hydrochloric acid solution. Be sure to inspect the buret tip for an "air pocket". The "air pocket" must be removed before starting the titration. Record the molarity of the hydrochloric acid solution and the initial buret reading on the laboratory Data Sheet in the section titled "Solubility Product Determination from Hydrochloric Acid Titration". Pipette 25.0 ml of the saturated calcium hydroxide solution that you initially took from the reagent bottle into a 125-ml Erlenmeyer flask. If you observe any solid residue, be sure not to transfer the residue inside the pipette. Add two or three drops of methyl orange indicator to the solution (methyl orange is yellow in basic solution and red in acidic solution, see Table 14.1). Titrate the saturated calcium hydroxide solution with 0.02 Molar hydrochloric acid solution until you observe a color change (yellow-to-red; if you add the hydrochloric acid slow enough, you will be able to stop the titration at the yellow-orange color change. The orange color is really what you want, however, the change from orange-red is so sudden (one drop or so) that most people will titrate to the red color. One drop "excess" hydrochloric acid will cause a negligible error – assuming that you stop at the first "permanent" appearance of red color.) Remember you want to stop the titration when a "permanent" color change is first observed, and not several ml later. When you think that the equivalence point (color change) is near, adjust the buret stopcock so that the addition of hydrochloric acid is dropwise. Record the final buret reading on the Data Sheet, and calculate the volume of hydrochloric used in the titration by difference.

Repeat the titration procedure a second time, again using 25 ml of the saturated calcium hydroxide solution. Again record the experimental data on the laboratory Data

Sheet. Once both titrations are finished, calculate the molarity of the hydroxide ion in the saturated calcium hydoxide solution by substituting the volume of saturated calcium hydroxide analyzed (25 ml), and the molarity and volume of the hydrochloric acid used in each titration, into Eqn. 22.7. The calculated hydroxide concentration is then inserted into Eqn. 22.2 (along with $[Ca^{2+}] = 0.5\ [OH^-]$) to get the numerical value of the solubility product. Calculate the solubility product of calcium hydroxide. Give the TA your value of K_{sp} for calcium hydroxide from the HCl titration. The TA will collect and compile the values for the entire class, and will post the information on the bulletin board outside of the laboratory. Go to the bulletin the next day, and enter the values for the entire class on your laboratory Data Sheet. Calculate the class mean and class standard deviation from the HCl titration.

CLEANING PROCEDURE FOR BURETTE

At the end of the experiment it is important to rinse the burette several times with distilled water to remove any titrant from the burette. Remove the stopcocks and rinse the stopcock assembly thoroughly. Reassemble the burette and rinse with deionized water. Invert the burette in the burette holder (that is, hang the burette upside down) with the stopcock open to dry. This will prevent the tip from from getting clogged with a solid. If the burette is not thoroughly cleaned the tip will likely become clogged with solid once the water has evaporated. Leave the stopcock in this configuration for the next laboratory session.

Note to Instructor – Barium hydroxide is another compound that can be used in the solubility product determination. See J. P. Reynolds, *J. Chem. Educ.*, **52**, 521-522 (1975).

DATA SHEET – EXPERIMENT 22

$$Ca(OH)_2 \begin{array}{c} \nearrow Ca^{2+}_{(aq)} \\ \searrow 2\,OH^-_{(aq)} \end{array}$$

Name: _____

Solubility Product Estimation using *pHydrion* Paper

1. Color of *pHydrion* strip: _____

2. pH of saturated solution based on color guide: _____

3. Hydronium ion concentration in saturated solution, M: _____

4. Hydroxide ion concentration in saturated solution, M: _____

5. Estimated solubility product for calcium hydroxide: _____

Solubility Product Determination from pH Electrode Measurement

6. pH of diluted calcium hydroxide solution: _____

7. Hydronium ion concentration in dilute solution, M: _____

8. Hydroxide ion concentration in diluted solution, M: _____

9. Hydroxide ion concentration in saturated solution, M: _____

10. Calculated solubility product of calcium hydroxide: _____

11. Literature value for solubility product of calcium hydroxide: _____

Solubility Product Determination from Hydrochloric Acid Titration

	Trial 1	Trial 2
12. Molarity of hydrochloric acid, M:	_____	_____
13. Initial buret reading before HCl addition, ml:	_____	_____
14. Final buret reading after HCl addition, ml:	_____	_____
15. Volume of hydrochloric acid used, ml:	_____	_____
16. [OH⁻] in saturated Ca(OH)₂ solution, M:	_____	_____
17. Calculated solubility product from titration:	_____	_____
18. Average value of solubility product from titration:	_____	
19. Literature value for solubility product of calcium hydroxide:	_____	

Class Data – Solubility Product Determination from Titration Data:

_____	_____	_____	_____
_____	_____	_____	_____
_____	_____	_____	_____
_____	_____	_____	_____
_____	_____	_____	_____

17. Class Mean Solubility Product: _____

18. Class Standard Deviation: _____

EXPERIMENT 23: REDOX TITRATION – STANDARDIZATION OF POTASSIUM PERMANGANATE SOLUTION

INTRODUCTION

Oxidation-reduction reactions, commonly referred to as redox reactions, involve the transfer of electrons between reactants. The corrosion of iron (rust formation) is

$$4 \ Fe_{(solid)} + 3 \ O_{2(gas)} \longrightarrow 2 \ Fe_2O_{3(solid)}$$

one such example of an oxidation-reduction reaction. Oxygen goes from an oxidation state of "0" (oxidation number of oxygen in O_2 is zero) to an oxidation state of "-2" (oxidation number of oxygen in Fe_2O_3 is -2). Each oxygen atom gains 2 electrons. Gain of electrons by a substance is called reduction. Iron, on the other hand, loses 3 electrons in going from metallic elemental iron (oxidation number of Fe is 0) to ferric oxide (oxidation state of Fe is +3 in Fe_2O_3). Loss of electrons by a substance is called oxidation. When one reactant loses electrons, another reactant must gain the lost electrons. For the overall chemical reaction, the loss and gain of electrons must be balanced. Many metals react directly with oxygen in air to form metal oxides. Other important examples of oxidation-reduction reactions are found in nonrechargeable alkaline and rechargeable nickel-cadmium (nicad) batteries, the lead-acid automotive battery, combustion processes, fuel cells, actions of bleaches, respirations of animals, and the biological ATP-ADP interconversion process. ATP-ADP interconversions are used to store energy during metabolism, and then later to release the stored energy as needed to drive nonspontaneous reactions in the body.

In Experiment 23 you will investigate a redox reaction using titration methods. There are two sets of titrations to be performed. In the first set of titrations you will "standardize" a solution of potassium permanganate ($KMnO_4$) by titrating it against a weighed amount of sodium oxalate ($Na_2C_2O_4$). Once the concentration of permanganate is known, the permanganate solution can then be used as a titrant to analyze unknown oxalate solutions. Permanganate ions (MnO_4^-) are strong oxidizing agents as they contain manganese in a very high oxidation state (+7). In an acidic solution MnO_4^- is able to oxidize oxalate ions ($C_2O_4^{2-}$) to carbon dioxide according to the following balanced chemical reaction:

$$2 \ MnO_4^-{}_{(aq)} + 5 \ C_2O_4^{2-}{}_{(aq)} + 16 \ H^+{}_{(aq)} \ \text{------->} \ 2 \ Mn^{2+}{}_{(aq)} + 10 \ CO_{2(gas)} + 8 \ H_2O_{(liq)}$$

The progress of the chemical can be followed through the color change of the ions. No indicator is needed for this titration. The MnO_4^- ion is purple, and the endpoint of the titration will be when the solution turns faint purple, and the purple color persists for at least 30 seconds. From the stoichiometric coefficients in the balanced chemical reaction, one finds that

$$\text{moles of MnO}_4^- = (2/5) \times \text{moles of C}_2\text{O}_4^{2-} \tag{23.1}$$

two moles of MnO_4^- are needed for every five moles of $C_2O_4^{2-}$. The substitution pattern for the number of moles of a chemical in a titration, is that the number of moles equals the mass of the compound divided by its molar mass, or the number of moles equals the product of the molarity of the compound times the volume (in liters). Which of the two substitution patterns used depends on the information known, or what we wish to calculate.

In the first set of titrations, the standardization titrations, we want to calculate the molarity of MnO_4^-, and we will know the volume of permanganate used (from the burette readings) and the mass of sodium oxalate ($Na_2C_2O_4$). Since there is one oxalate ion per sodium oxalate, the number of moles of oxalate ion will be equal to the number of moles of sodium oxalate. The substitution pattern for the standardization titration is then to replace the moles of MnO_4^- in Eqn. 23.1 by molarity of MnO_4^- × volume of MnO_4^- (in liters) and the number of moles of oxalate by mass of sodium oxalate divided by molar mass of sodium oxalate (Molar mass of $Na_2C_2O_4$ = 134.00 grams/mole). After performing these substitutions, we get

$$\text{molarity of MnO}_4^- \times \text{volume of MnO}_4^- = (2.5) \times (\text{mass Na}_2\text{C}_2\text{O}_4/ \text{molar mass Na}_2\text{C}_2\text{O}_4) \tag{23.2}$$

For the standardization reaction, every quantity in Eqn. 23.2 is known, except for the Molarity of MnO_4^-, which we wish to calculate. Unlike some of the acid-base experiments that were performed earlier in the semester, it is not possible to accurately know the concentration of MnO_4^- in the solution by dissolving a known weight of $KMnO_4$ in a known volume of distilled water. It is very difficult to obtain potassium permanagate in an entirely pure state; it is generally contaminated by traces of manganese dioxide, MnO_2. The presence of MnO_2 is undesired, in that it catalyzes the autodecomposition of permanganate ion on standing. For accurate analytical results, the accepted practice is to experimentally determine the concentration of the permanganate solution just prior to its use.

In the second set of titrations, we want to determine the molarity of an unknown oxalate ion solution, using MnO_4^- again as the titrant. For the second set of titrations, the molarity of MnO_4^- is known. That was the purpose of doing the standardization titration. Since we want to calculate the molarity of oxalate ion, the logical substitution pattern for the number of moles of $C_2O_4^{2-}$ in Eqn. 23.1 would be to replace the quantity with the product of the molarity of $C_2O_4^{2-}$ × volume of $C_2O_4^{2-}$ (in liters). This substitution yields

$$\text{molarity of MnO}_4^- \times \text{volume of MnO}_4^- =$$

$$(2/5) \times (\text{molarity of } C_2O_4^{2-} \times \text{volume of } C_2O_4^{2-})$$

(23.3)

Every quantity in Eqn 21.3 is known, except for the molarity of $C_2O_4^{2-}$, which we are asked to experimentally determine. (The volume of $C_2O_4^{2-}$ is 25.0 ml, the amount of unknown solution that was transferred to the Erlenmeyer flask at the beginning of the titration.)

PROCEDURE FOR STANDARDIZATION OF POTASSIUM PERMANGANATE SOLUTION

Weigh a 250-ml Erlenmeyer flask, and record the weight on the laboratory Data Sheet just opposite "mass of empty flask". Add approximately 0.2 to 0.3 grams of sodium oxalate to the Erlenmeyer flask. Reweigh the flask, and record the "mass of flask + sodium oxalate" on the Data Sheet.

CAUTION – SODIUM OXALATE IS A POISON – DO NOT INGEST

Add 100 ml of distilled water and 30 ml of 3 Molar sulfuric acid. Remember to add acid to water (and not the reverse). Gently swirl (stir) until all of the sodium oxalate has dissolved.

CAUTION – SULFURIC ACID IS A STRONG ACID AND IS CAUSTIC. IT CAN CAUSE BURNS IF LEFT IN CONTACT WITH THE SKIN. IF ANY SULFURIC ACID GETS ON YOUR SKIN IMMEDIATELY RINSE WITH PLENTY OF COOL WATER FOR SEVERAL MINUTES. INFORM THE TEACHING ASSISTANT IF YOU SPILL ANY SULFURIC ACID ON THE BENCHTOP.

Fill a buret with $KMnO_4$ solution. **INSPECT THE TIP OF THE BURETTE TO MAKE SURE THAT THERE IS NO "AIR POCKET" THAT COULD BECOME DISLODGED DURING THE COURSE OF THE TITRATION.** If you do notice an air pocket, open the stopcock slightly and allow a few ml of titrant to drain. This will generally eliminate the air pocket. If not, gently tap on the tip of the burette while the titrant is draining. This often helps. Before you begin the titration the **air pocket must be removed**. Otherwise, the titrant will displace the air procket during the titration, and not all of the titrant used will be delivered to the titration flask. Some will remain in the space that was once occupied by the air pocket. Now record the initial buret reading on the Data Sheet in the space labeled "Initial buret reading". For a dark or deeply colored solution the meniscus may not be clearly visible. The liquid level may appear instead to go horizontally across, in which case this could be used as means to measure the initial and final buret readings. Add about 15 ml of the permanganate solution to your sodium oxalate solution, swirl and then allow to stand until the purple color has disappeared. If you have problems, repeat using less $KMnO_4$.

CAUTION – POTASSIUM PERMANGANATE WILL STAIN SKIN ON CONTACT.

After the purple color has disappeared, you will need to heat the solution gently to about 50 to 60 °C (at this temperature the flask will feel hot but not uncomfortable to hold). At this point, remove the heat and resume the titration. You should continue the titration until a pale purple color persists for at least 30 seconds. Add the last 0.5 to 1.0 ml of $KMnO_4$ dropwise with particular care to allow each drop to become decolorized. Record the final buret reading of $KMnO_4$ added on the Data Sheet. Repeat the entire procedure a second time, and record the second set of experimental data on the Data Sheet under the column heading Trial 2. Calculate the concentration of the potassium permanganate solution and the value to your TA. The TA will collect and compile the values for the entire class, and will post the information on the bulletin board outside the laboratory.

PROCEDURE FOR DETERMINATION OF CONCENTRATION OF UNKNOWN OXALATE SOLUTION

Your TA will give you a solution of unknown oxalate ion concentration. Transfer by pipet or buret exactly 25.00 ml of the unknown oxalate solution into a clean 250-ml Erlenmeyer flask. Add 75 ml of distilled water and 20 ml of 3 Molar H_2SO_4. Mix the chemicals by swirling and repeat the titration procedure as described in the preceding standardization instructions. Record the initial and final buret readings for $KMnO_4$ on the Data Sheet. Repeat the process a second time, and record the data for the second trial on the Data Sheet.

CLEANING PROCEDURE FOR BURETTE

At the end of the experiment it is important to rinse the burette several times with distilled water to remove any titrant from the burette. Remove the stopcocks and rinse the stopcock assembly thoroughly. Reassemble the burette and rinse with deionized water. Invert the burette in the burette holder (that is, hang the burette upside down) with the stopcock open to dry. This will prevent the tip from getting clogged with a solid. If the burette is not thoroughly cleaned the tip will likely become clogged with solid once the water has evaporated. Leave the stopcock in this configuration for the next laboratory session.

Complete the Data Sheet by performing the calculations discussed in the Introduction. Obtain the class data for the standardization titrations, and calculate the mean and standard deviation for the potassium permanganate solution.

DATA SHEET – EXPERIMENT 23

Name: _____

Standardization of KMnO$_4$	Trial 1	Trial 2
1. Mass of flask + sodium oxalate, g:	_____	_____
2. Mass of empty flask, g:	_____	_____
3. Mass of sodium oxalate, g:	_____	_____
4. Initial buret reading of KMnO$_4$, ml:	_____	_____
5. Final buret reading of KMnO$_4$, ml:	_____	_____
6. Volume of KMnO$_4$ used, in ml:	_____	_____
7. Volume of KMnO$_4$ used, in l:	_____	_____
8. Molarity of MnO$_4^-$ solution, M:	_____	_____
9. Average molarity of MnO$_4^-$ solution:	_____	

Determination of Concentration of Unknown Oxalate Solution

	Trial 1	Trial 2
10. Volume of unknown oxalate solution, l:	_____	_____
11. Initial buret reading of KMnO$_4$, ml:	_____	_____
12. Final buret reading of KMnO$_4$, ml:	_____	_____
13. Volume of KMnO$_4$ used, in ml:	_____	_____
14. Volume of KMnO$_4$ used, in l:	_____	_____
15. Molarity of unknown oxalate solution, M:	_____	_____
16. Average concentration of oxalate solution, M:	_____	

Class Data – Molarity of Potassium Permanganate Solution:

_____	_____	_____	_____
_____	_____	_____	_____
_____	_____	_____	_____
_____	_____	_____	_____
_____	_____	_____	_____

17. Class Mean Molarity of $KMnO_4$: _____

18. Class Standard Deviation: _____

EXPERIMENT 24: DETERMINATION OF THE ACTIVITY SERIES FOR METALS

INTRODUCTION

Extensive studies with many metals have led to the development of a metal activity series, which is a ranking of the relative reactivity of metals in displacement and other kinds of oxidation-reduction reactions. The most reactive metals (like Li, K, Ba) appear at the top of the series, and are very powerful reducing agents and readily form cations. Metals near the bottom of the series (Au, Ag) are poor reducing agents. Their cations (Au^+, Ag^+), however, are powerful oxidizing agents that readily react to form the free metal.

An element higher in the activity series will displace an element below it in the series from its components. For example, metallic zinc displaces copper from a Cu^{2+} solution and nickel form a Ni^{2+} solution

$$Zn_{(solid)} + Cu^{2+}_{(aq)} \text{--------}> Zn^{2+}_{(aq)} + Cu_{(solid)}$$
$$Zn_{(solid)} + Ni^{2+}_{(aq)} \text{--------}> Zn^{2+}_{(aq)} + Ni_{(solid)}$$

This means that zinc metal must lie above copper metal and nickel metal in the activity series.

In today's laboratory experiment you will construct a partial metal activity series by studying which cations a given metal is able to displace from solution. The series of half-reactions that will be studied are listed below:

$$H^+_{(aq)} + e^- \text{-------}> H_{2(gas)}$$
$$Na^+_{(aq)} + e^- \text{--------}> Na_{(solid)}$$
$$Al^{3+}_{(aq)} + 3\ e^- \text{--------}> Al_{(solid)}$$
$$Fe^{3+}_{(aq)} + 3\ e^- \text{--------}> Fe_{(solid)}$$
$$Ni^{2+}_{(aq)} + 2\ e^- \text{-------}> Ni_{(solid)}$$
$$Cu^{2+}_{(aq)} + 2\ e^- \text{-------}> Cu_{(solid)}$$
$$Zn^{2+}_{(aq)} + 2\ e^- \text{--------}> Zn_{(solid)}$$

The half-reactions are not listed in order of reactivity, but rather in terms of increasing atomic number. In deciding whether or not a reaction has occurred, look for color changes (both the solution and the metal), gas bubbles and the appearance of new substances.

EXPERIMENTAL PROCEDURE – GROUP DEMONSTRATIONS

It will take too long for you to perform all of the experiments necessary to establish the activity series. This experiment is to be done as a group experiment where each person (or group) is responsible for setting up and conducting one of the following set of experiments for class display. Sequentially each group should perform their task as the other students watch. Some of the reactions may be immediate, but others may take several minutes for some observable change to occur. Each display should be observed by each student a second time after a half-hour has past. That is, all students will observe each demonstration as it is first performed, and then again, about a half-hour later. Each reaction should be carefully labeled with a piece of paper on the laboratory bench top so that all persons will be able to identify the respective reaction when they make their second tour around the laboratory room. The labels need to clearly indicate the reaction, for example, "$Zn + Cu(NO_3)_2$ " or "$HCl + Fe$".

Group 1:

Place about 2 ml of each of the following in different test tubes (all solutions should be at least 1 Molar):

HCl $Fe(NO_3)_3$ $Ni(NO_3)_2$ $Zn(NO_3)_2$ $Cu(NO_3)_2$ $NaNO_3$

Add a clean piece or strip of copper to each test tube, swirl and record your observations with the entire class watching.

Group 2:

Place about 2 ml of each of the following in different test tubes (all solutions should be at least 1 Molar):

HCl $Fe(NO_3)_3$ $Ni(NO_3)_2$ $Zn(NO_3)_2$ $Cu(NO_3)_2$ $NaNO_3$

Add a clean piece or strip of iron to each test tube, swirl and record your observations with the entire class watching.

Group 3:

Place about 2 ml of each of the following in different test tubes (all solutions should be at least 1 Molar):

HCl $Fe(NO_3)_3$ $Ni(NO_3)_2$ $Zn(NO_3)_2$ $Cu(NO_3)_2$ $NaNO_3$

Add a clean piece or strip of nickel to each test tube, swirl and record your observations with the entire class watching.

Group 4:

Place about 2 ml of each of the following in different test tubes (all solutions should be at least 1 Molar):

$HCl \qquad Fe(NO_3)_3 \qquad Ni(NO_3)_2 \qquad Zn(NO_3)_2 \qquad Cu(NO_3)_2 \qquad NaNO_3$

Add a clean piece or strip of zinc to each test tube, swirl and record your observations with the entire class watching.

Group 5:

Place about 2 ml of each of the following in different test tubes (all solutions should be at least 1 Molar):

$HCl \qquad Fe(NO_3)_3 \qquad Ni(NO_3)_2 \qquad Zn(NO_3)_2 \qquad Cu(NO_3)_2 \qquad NaNO_3$

Add a clean piece or strip of aluminum to each test tube, swirl and record your observations with the entire class watching. You will have to gently sand the aluminum with sandpaper or an emery board to remove any oxide coating that might have formed.

Additional Observation:

When sodium metal is added to water it reacts vigorously/violently. Water is a weak acid, having $[H_3O^+] = 1 \times 10^{-7}$. This piece of information combined with the observations that you made in regards to the sodium nitrate solutions should allow you to place correctly sodium metal on the metal activity scale.

DATA SHEET – EXPERIMENT 24

$$Cu^{2+}_{(aq)} + 2\ e^{-} \longrightarrow Cu_{(solid)}$$

Name: _____

Observations Concerning Metal Activity Group Demonstrations

1. **Group 1:**

   Describe the reaction of copper metal with:

   HCl: _____

   $Fe(NO_3)_3$: _____

   $Ni(NO_3)_2$: _____

   $Zn(NO_3)_2$: _____

   $Cu(NO_3)_2$: _____

   $NaNO_3$: _____

2. **Group 2:**

   Describe the reaction of iron metal with:

   HCl: _____

   $Fe(NO_3)_3$: _____

   $Ni(NO_3)_2$: _____

   $Zn(NO_3)_2$: _____

   $Cu(NO_3)_2$: _____

   $NaNO_3$: _____

3. **Group 3:**

   Describe the reaction of nickel metal with:

   HCl: _____

   $Fe(NO_3)_3$: _____

   $Ni(NO_3)_2$: _____

   $Zn(NO_3)_2$: _____

   $Cu(NO_3)_2$: _____

   $NaNO_3$: _____

4. **Group 4:**

Describe the reaction of zinc metal with:

HCl: _____

$Fe(NO_3)_3$: _____

$Ni(NO_3)_2$: _____

$Zn(NO_3)_2$: _____

$Cu(NO_3)_2$: _____

$NaNO_3$: _____

5. **Group 5:**

Describe the reaction of aluminum metal with:

HCl: _____

$Fe(NO_3)_3$: _____

$Ni(NO_3)_2$: _____

$Zn(NO_3)_2$: _____

$Cu(NO_3)_2$: _____

$NaNO_3$: _____

Order of Metal Activity

Label the 6 half-reactions below in order of decreasing tendency to proceed in the indicated direction. Place "1" besides the least reactive metal, and "6' besides the most reactive metal. The H^+/H half-reaction is also included in the list.

_____ $H^+_{(aq)} + e^- \text{-------> } H_{2(gas)}$

_____ $Na^+_{(aq)} + e^- \text{--------> } Na_{(solid)}$

_____ $Al^{3+}_{(aq)} + 3 e^- \text{--------> } Al_{(solid)}$

_____ $Fe^{3+}_{(aq)} + 3 e^- \text{--------> } Fe_{(solid)}$

_____ $Ni^{2+}_{(aq)} + 2 e^- \text{-------> } Ni_{(solid)}$

_____ $Cu^{2+}_{(aq)} + 2 e^- \text{-------> } Cu_{(solid)}$

_____ $Zn^{2+}_{(aq)} + 2 e^- \text{--------> } Zn_{(solid)}$

Post Laboratory Exercise

Given below is the activity reaction order for an additional five metals:

$Mn_{(metal)}$ ------> $Mn^{2+}_{(aq)} + 2\ e^-$

$Co_{(metal)}$ ------> $Co^{2+}_{(aq)} + 2\ e^-$

$Sn_{(metal)}$ ------> $Sn^{2+}_{(aq)} + 2\ e^-$

$Pb_{(metal)}$ ------> $Pb^{2+}_{(aq)} + 2\ e^-$

$Ag_{(metal)}$ ------> $Ag^{+}_{(aq)} + e^-$

with the most reactive metal at the top, and the least reactive metal at the bottom. For each of the following demonstrations, circle the cation(s) that the metal is (are) expected to react with.

Demonstration 1:

A piece of lead metal is placed in 2 ml of each of the following solutions:

(a) Co^{2+} (b) Sn^{2+} (c) Ag^+ (d) Mn^{2+}

Demonstration 2:

A piece of tin metal is placed in 2 ml of each of the following solutions:

(a) Co^{2+} (b) Pb^{2+} (c) Ag^+ (d) Mn^{2+}

Demonstration 3:

A piece of silver metal is placed in 2 ml of each of the following solutions:

(a) Co^{2+} (b) Sn^{2+} (c) Pb^{2+} (d) Mn^{2+}

Demonstration 4:

A piece of cobalt metal is placed in 2 ml of each of the following solutions:

(a) Pb^{2+} (b) Sn^{2+} (c) Ag^+ (d) Mn^{2+}

Demonstration 5:

A piece of manganese metal is placed in 2 ml of each of the following solutions:

(a) Co^{2+} (b) Sn^{2+} (c) Ag^+ (d) Pb^{2+}

EXPERIMENT 25: ELECTROCHEMISTRY – VERIFICATION OF THE NERNST EQUATION

INTRODUCTION

Oxidation-reduction reactions involve the transfer of electrons from one chemical species to another. The energy released during a spontaneous oxidation-reduction reaction can be used to perform electrical work. In today's laboratory experiment you will construct several electrochemical cells, and examine how the measured cell voltage depends upon the reactant and product concentrations. One of the electrochemical cells that will be studied involves the reaction between zinc metal and copper(II) ions. The electrochemical cell is constructed by placing a solution of 0.100 Molar copper(II) nitrate (or copper(II) sulfate) in one beaker and a solution of 0.100 Molar zinc nitrate (or zinc sulfate) in a second beaker. A zinc metal strip is placed in the solution of zinc nitrate, and a strip of copper metal is placed in the copper(II) nitrate solution. Both metal strips are then connected to a potentiometer (volt-meter) using a platinum wire and alligator clips as shown in Figure 25.1. The two solutions are also joined by a inverted U-shaped tube (called a salt-bridge) that contains an electrolyte solution, such as $NaNO_{3(aq)}$, whose ions will not interact with other ions in the cell, or with the metal electrode materials.

In the Cu-Zn electrochemical cell, electrons become available as zinc metal is

$$Zn_{(metal)} \;\; ----> \;\; Zn^{2+}_{(aq)} + 2\,e^-$$

oxidized at the anion. The electrons flow through the external circuit to the cathode, where they are consumed by Cu^{2+}:

$$Cu^{2+}_{(aq)} + 2\,e^- \;\; ------> \;\; Cu_{(metal)}$$

in the solution. The net result is the flow of electrons from the anode (Zn metal electrode) to the cathode (Cu metal electrode). As the chemical reaction proceeds the Zn^{2+} ion molar concentration increases, while the Cu^{2+} ion concentration decreases. Sodium ions from the salt bridge (which is the electrolyte $NaNO_3$ in this example) migrate to the Cu^{2+} solution to replace the positive charge lost as the result of Cu reduction. Nitrate ions in the salt bridge migrate in the opposite direction, to the Zn^{2+} solution to counterbalance the built up of Zn^{2+} positive charge that results when zinc metal is oxidized. The flow of ions from the salt bridge maintains the condition of electrical neutrality in all three solutions.

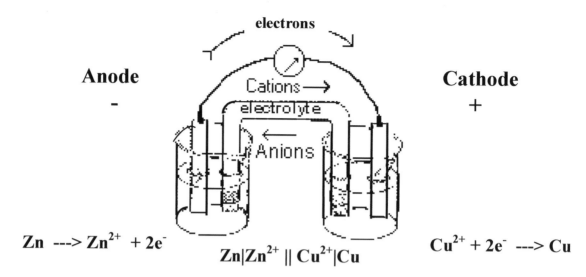

Anode

-

Cathode

+

$Zn \longrightarrow Zn^{2+} + 2e^-$

$Zn|Zn^{2+} \parallel Cu^{2+}|Cu$

$Cu^{2+} + 2e^- \longrightarrow Cu$

Figure 25.1. Cu-Zn electrochemical cell

Chemical equilibrium are mathematically expressed in terms of an equilibrium constant. Half-reactions, like the two used in the electrochemical cell depicted in Figure 25.1, are described by the Nernst equation

Cathode: $Cu^{2+}_{(aq)} + 2e^- \text{-------}> Cu_{(metal)}$

$$E_{cathode} = E^o_{Cu(II)/Cu} - (2.303 \ RT/n \ F) \log (1/[Cu^{2+}]) \qquad (25.1)$$

$$E_{cathode} = E^o_{Cu(II)/Cu} - (0.0591/2) \log (1/[Cu^{2+}]) \qquad (25.2)$$

Anode: $Zn^{2+}_{(aq)} + 2e^- \text{-------}> Zn_{(metal)}$

$$E_{anode} = E^o_{Zn(II)/Zn} - (2.303 \ RT/n \ F) \log (1/[Zn^{2+}]) \qquad (25.3)$$

$$E_{anode} = E^o_{Zn(II)/Zn} - (0.0591/2) \log (1/[Zn^{2+}]) \qquad (25.4)$$

where E^o denotes the standard reduction for the half-reaction couple, T is the absolute Kelvin temperature, R and F refer to the universal ideal gas law constant (8.314 J mol^{-1} K^{-1}) and Faraday's constant (96,485 coulombs mole^{-1}), respectively, and n is the number of electrons transferred in each half-reaction. Standard reduction potentials correspond to 1 Molar concentration. The logarithmic term allows one to calculate the electrode potentials at other concentrations of Cu^{2+} and Zn^{2+} in the above electrochemical cell. At 298.15 K, the coefficient in front of the logarithmic value equals (2.303 RT/F) = 0.0591 volt.

The potentiometer that is connected to the copper and zinc metal strips in Figure 25.1 measures:

$$E_{cell} = E_{cathode} - E_{anode} \qquad (25.5)$$

$$E_{cell} = E°_{Cu(II)/Cu} - E°_{Zn(II)/Zn} - (0.0591/2) \log (1/[Cu^{2+}])$$
$$+ (0.0591/2) \log (1/[Zn^{2+}]) \qquad (25.6)$$

the potential difference between the cathode and anode electrodes. In today's laboratory experiment you will constant this electrochemical verify that the Nernst equation is obeyed for several different molar concentrations of Cu^{2+} and Zn^{2+}. The short-hand cell notation for writing the cells that you will construct is:

$$Zn \mid Zn^{2+} \text{ (x Molar)} \parallel Cu^{2+} \text{ (x Molar)} \mid Cu$$

By convention the anode half-reaction is given on the left hand side of "$\parallel$", and the cathode half-reaction appears to the right. The double line "$\parallel$" indicates the salt bridge, and the single line "$\mid$" separates the metal electrode from the dissolved ions in the solution.

One way to verify that the Nernst equation is obeyed would be to compare point-by-point the measured cell voltage for each Cu and Zn concentration studied to the values calculated from Eqn. 25.6. The standard reduction potentials for the Cu(II)/Cu - couple, $E°_{Cu(II)/Cu} = 0.339$ volt, and for the Zn(II)/Zn - couple, $E°_{Zn(II)/Zn} = -0.762$ volt, are both known. Alternatively, Eqn. 25.6 can be algebraically manipulated into a linear form:

$$E_{cell} = E°_{Cu(II)/Cu} - E°_{Zn(II)/Zn} + (0.0591/2) \log ([Cu^{2+}]/[Zn^{2+}]) \qquad (25.7)$$

If the Nernst equation is obeyed, a plot of E_{cell} (y-axis) versus $\log ([Cu^{2+}]/[Zn^{2+}])$ (x-axis) should be linear. A linear least squares analysis will be performed to determine whether or not the graph is linear. Remember a correlation coefficient of $r = 1$ would indicate a "perfect" fit of the experimental data points to the line. Does the verification entail anything else besides simply observing whether or not the graph is linear? In other words, are any restrictions placed on the linear relationship? Yes, if the Nernst equation is obeyed the slope of the line should be (0.0591/2) volts (or 29.5 millivolts), and the y-intercept should equal 1.101 volts. Compare the numerical values that you obtain for the slope and intercept from the linear least squares analysis to the theoretical values.

After the Nernst equation is verified, it will be used in the determination of the standard reduction potential for the following half-reaction:

$$Ni^{2+}_{(aq)} + 2 e^- \text{ --------> } Ni_{(metal)}$$

$$E_{electrode} = E°_{Ni(II)/Ni} - (2.303 \, RT/n \, F) \log (1/[Ni^{2+}]) \qquad (25.8)$$

The standard electrode potential for this half-reaction is denoted as $E°_{Ni(II)/Ni}$. This part of the experiment involves constructing two additional electrochemical cells. The first cell is

$$Ni \mid Ni^{2+} \text{ (0.10 Molar)} \parallel Cu^{2+} \text{ (0.10 Molar)} \mid Cu$$

constructed by replacing the zinc metal and 0.10 Molar $Zn(NO_3)_2$ (or $ZnSO_4$) with nickel metal and 0.10 Molar $Ni(NO_3)_2$ solution. The Ni strip is the anode electrode for this electrochemical cell. In the second electrochemical cell:

$$Zn \mid Zn^{2+} \text{ (0.10 Molar)} \parallel Ni^{2+} \text{ (0.10 Molar)} \mid Ni$$

the copper metal strip and 0.10 Molar $Cu(NO_3)_2$ (or $CuSO_4$) solution is replaced with a nickel metal strip and 0.10 Molar $Ni(NO_3)_2$ solution. E_{cell} expressions for the Ni-Cu and Zn-Ni electrochemical cells are obtained by substituting the respective E_{cathod} and E_{anode} Nernst equations (see Eqns. 25.2, 25.4 and/or 25.8) into $E_{cell} = E_{cathode} - E_{anode}$. Be sure that the Nernst equations are substituted correctly. The cell notation identifies the anode and cathode for each electrochemical cell.

EXPERIMENTAL PROCEDURE FOR VERIFICATION OF NERNST EQUATION

You will need to construct several electrochemical cells in which the Cu^{2+} and Zn^{2+} ion concentrations are varied. Prepare solutions that contain the following concentrations of Cu^{2+}: 0.100 Molar, 0.0100 Molar and 0.00100 Molar. The solutions can be prepared quickly by serial dilution. To prepare the 0.0100 Molar solution, you will need to dilute 10 ml of the 0.10 Molar solution with 90 ml of deionized water. Now dilute 10 ml of the 0.0100 Molar solution that was just prepared with 90 ml of deionized water. The 0.00100 Molar Cu^{2+} solution is prepared by this later dilution. Similarly you will need to prepare solutions that contain the following concentrations of Zn^{2+}: 0.100 Molar, 0.0100 Molar and 0.00100 Molar.

The different electrochemical cells are constructed using an ice cube tray. Fill one ice cube cell with a different solution as shown in Figure 25.2. (Be sure to the label the ice cube cells so that you will quickly know what solution is in each cell.) Take the strips of zinc and copper metal strips, and sand gently to remove any oxide or corrosion film. Now place the zinc metal strip into the Zn^{2+} cell that you wish to use, and the copper metal strip goes into the Cu^{2+} cell that you want to study. Attach the black alligator clip to the zinc metal strip and the red alligator clip to the copper metal. For the Pasco Voltage/Current Sensor the red wire is (+) and the black wire is (-). Now take a piece of filter paper, which has been soaked in 0.10 Molar sodium nitrate solution, and connect the two cells. Place the sodium nitrate soaked filter paper only when you get ready to actually make the measurement. Remove the filter paper immediately after the measurement is made. You want the filter paper to connect the two solutions only during the brief time that it takes to measure the cell voltage, otherwise the Cu^{2+} and Zn^{2+} solutions will be contaminated with each other. The Cu^{2+} solution will travel up the filter paper (called wicking), and will eventually migrate over to the Zn^{2+} solution. The Zn^{2+}

solution will do likewise. Cross contamination can be minimized/eliminated by limiting the time that the two solutions are connected *via* the $NaNO_3$ soaked filter-paper salt bridge. To even reduce the chance for cross contamination even further, you may want to use a new piece of $NaNO_3$ soaked filter paper for each measurement.

0.100 M Zn^{2+}	0.100 M Cu^{2+}
0.0100 Molar Zn^{2+}	0.100 Molar Cu^{2+}
0.00100 Molar Zn^{2+}	0.100 Molar Cu^{2+}
0.100 Molar Zn^{2+}	0.0100 Molar Cu^{2+}
0.100 Molar Zn^{2+}	0.00100 Molar Cu^{2+}
0.010 M Zn^{2+}	0.010 M Cu^{2+}

Figure 25.2. Arrangement of solutions in the ice-cube tray
electrochemical set-up for verification of Nernst equation.

Measure the voltages for the following electrochemical cells:

Cell 1:   Zn | 0.100 Molar Zn^{2+} || 0.100 Molar Cu^{2+} | Cu

Cell 2:   Zn | 0.0100 Molar Zn^{2+} || 0.100 Molar Cu^{2+} | Cu

Cell 3:   Zn | 0.00100 Molar Zn^{2+} || 0.100 Molar Cu^{2+} | Cu

Cell 4:   Zn | 0.100 Molar Zn^{2+} || 0.0100 Molar Cu^{2+} | Cu

Cell 5:   Zn | 0.100 Molar Zn^{2+} || 0.00100 Molar Cu^{2+} | Cu

Cell 6:   Zn | 0.0100 Molar Zn^{2+} || 0.0100 Molar Cu^{2+} | Cu and record the experimental values on the laboratory Data Sheet in the section titled "Experimental Verification of the Nernst Equation". Make sure to gently sand the metal strips between each reading to remove any oxide or corrosion film. Also, remember to follow the cautionary note regarding the filter paper salt bridge. Two of the cells should have the same cell voltage. Do you know which two?

After all of the experimental data is measured, perform a linear least squares analysis using the built-in Pasco software. Enter the numerical values of $\log ([Cu^{2+}]/[Zn^{2+}])$ in the x-column of the Pasco table and corresponding E values in the y-column. Once all of the values have been entered, **mouse click on the Fit button, scroll down to linear fit, and mouse click**. A window should appear on the computer screen with all of the relevant statistical information. Record the slope and intercept of the line, along with the correlation coefficient, on your laboratory Data Sheet.

DETERMINATION OF STANDARD REDUCTION POTENTIAL FOR THE Ni^{2+}/Ni HALF REACTION

To measure the standard reduction potential for the Ni^{2+}/Ni half-reaction, fill four of the ice cube trays with 0.10 Molar Ni^{2+}, 0.10 Molar Cu^{2+} and 0.10 Molar Zn^{2+} solution as shown in Figure 25.3. Gently sand the copper and nickel metal strips with sandpaper. Place the copper strip in the copper solution and the nickel strip in the nickel solution. Connect the alligator clips – Remember for the Pasco Voltage/Current Sensor the red wire is (+) and the black wire is (-). Black wire to nickel metal and red wire to copper metal. Now take a piece of filter paper, which has been soaked in 0.10 Molar sodium nitrate solution, and connect the two cells. Measure the cell voltage and record the value on the laboratory Data Sheet in the section titled "Standard Reduction Potential Measurement for Ni^{2+}/Ni Couple". Repeat the procedure for the Zn-Ni electrochemical cell – this time the black wire goes to zinc metal and the red wire goes to the nickel metal. Remember to sand the zinc and nickel strips. Record the measured value on the Data Sheet in the appropriate place. Now calculate the standard reduction potential for the Ni^{2+}/Ni half reaction based on the measured cell voltages for both the Zn-Ni and Ni-Cu electrochemical cells. For this calculation you will need: $E^{\circ}_{Cu(II)/Cu} = 0.339$ volt and $E^{\circ}_{Zn(II)/Zn} = -0.762$ volt,

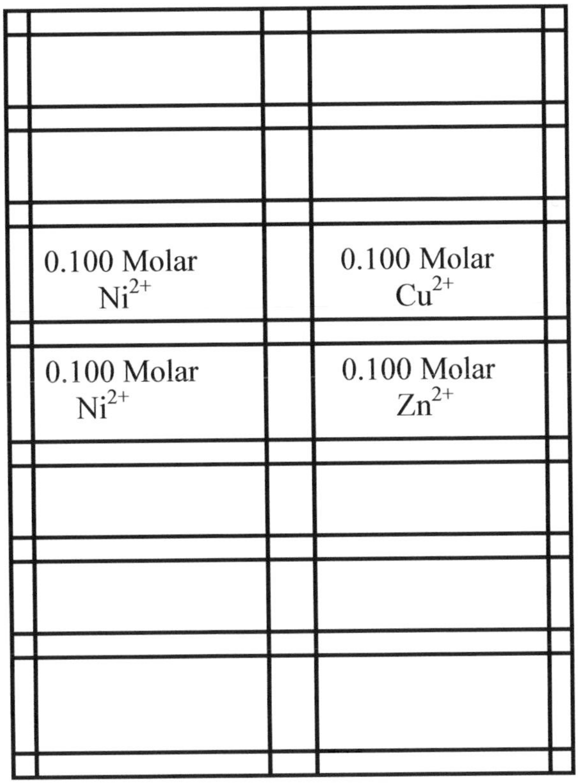

Figure 25.3. Arrangement of solutions in the ice-cube tray electrochemical set-up for determination of the standard reduction potential for Ni^{2+}/Ni half-reaction.

DATA SHEET – EXPERIMENT 25

Electrochemical cell

Name: _____

Experimental Verification of the Nernst Equation

1. Measured voltages for the following electrochemical cells:

Voltage

Cell 1:	$Zn \mid 0.100$ Molar $Zn^{2+} \parallel 0.100$ Molar $Cu^{2+} \mid Cu$:	_____
Cell 2:	$Zn \mid 0.0100$ Molar $Zn^{2+} \parallel 0.100$ Molar $Cu^{2+} \mid Cu$:	_____
Cell 3:	$Zn \mid 0.00100$ Molar $Zn^{2+} \parallel 0.100$ Molar $Cu^{2+} \mid Cu$:	_____
Cell 4:	$Zn \mid 0.100$ Molar $Zn^{2+} \parallel 0.0100$ Molar $Cu^{2+} \mid Cu$:	_____
Cell 5:	$Zn \mid 0.100$ Molar $Zn^{2+} \parallel 0.00100$ Molar $Cu^{2+} \mid Cu$:	_____
Cell 6:	$Zn \mid 0.0100$ Molar $Zn^{2+} \parallel 0.0100$ Molar $Cu^{2+} \mid Cu$:	_____

2. Value of slope of E_{cell} versus log ($[Cu^{2+}]/[Zn^{2+}]$) graph: _____

3. Theoretical value of the slope for this graph: _____

4. Value of y-intercept of E_{cell} versus log ($[Cu^{2+}]/[Zn^{2+}]$) graph: _____

5. Theoretical value of y-intercept for this graph: _____

6. Correlation coefficient of E_{cell} versus log ($[Cu^{2+}]/[Zn^{2+}]$) graph: _____

Standard Reduction Potential Measurement for Ni²⁺/Ni Couple

7. Measured voltage for the Ni-Cu electrochemical cell, v: _____

8. Measured voltage for the Zn-Ni electrochemical cell, v: _____

9. Calculated value of $E°_{Ni(II)/Ni}$ based on Ni-Cu cell, v: _____

10. Calculated value of $E°_{Ni(II)/Ni}$ based on Zn-Ni cell, v: _____

11. Discuss how your measured values compare to the literature value of $E^o_{Ni(II)/Ni}$ = -0.236 volt.

EXPERIMENT 26. SPECTROSCOPIC DETERMINATION OF THE EQUILIBRIUM CONSTANT FOR AN ACID-BASE INDICATOR

INTRODUCTION

Many dissolved chemical substances, both synthetic and naturally occurring, display colors that depend upon the hydronium ion concentration of the dissolving solution. The pH indicators that were used in Experiment 14 to locate the equivalence point in the acetic acid + sodium hydroxide titration are such examples. An acid-base indicator is a weak organic acid or a weak organic base whose protonated form differs in color from its unprotonated conjugate base. The chemical equilibrium involving the protonated, Hin, and unprotonated, In⁻, forms of the indicator

$$HIn_{(aq)} + H_2O \quad <\text{---------}> \quad H_3O^+_{(aq)} + In^-_{(aq)}$$

are described by

$$K_{HIn} = \frac{[H_3O^+][In^-]}{[HIn]} \tag{26.1}$$

an equilibrium constant. In today's laboratory experiment you will measure the equilibrium constant for bromothymol blue. Both forms of the indicator absorb in the visible spectral region; *e.g.*, protonated form appears yellow in color (absorption maxima near 450 nm) whereas the unprotonated form is blue (absorption near 550 nm).

Determination of the equilibrium constant is relatively straightforward, provided that one can substitute actual numerical values for the three molar concentration on the right-hand side of Eqn. 26.1. The measurement of $[H_3O^+]$ can be made with a pH electrode ($[H_3O^+] = 10^{-pH}$). The other two concentrations can be obtained through spectrophotometric absorption measurements. The experimental methodology involves first calculating the molar absorptivity, ε, of the protonated and unprotonated forms from measured absorption data for known bromothymol concentrations under both strongly acidic and strongly basic conditions. The molar absorptivities, once determined, are then used to determine equilibrium molar concentrations of HIn and In needed in Eqn. 26.1. For the equilibrium constant determination, the pH of the solutions must be in the color transition interval; otherwise, there would be too much experimental uncertainty associated with at least one (and possibly both) of the molar concentrations.

There is a fairly clever way of reducing the number of experimental measurements. Can you think of the method? Rather than determining actual experimental values for [HIn] and [In⁻], lets approach the problem from the standpoint "is there any special set of experimental conditions under which both [HIn] and [In⁻] could be eliminated from Eqn. 26.1. Yes, if the two values are equal, then $K_{HIn} = [H_3O^+]$. In the case of the determination of the acid dissociation constant of a weak monoprotoic weak acid by pH titration curve measurements (see Experiment 15), you used the pH of the solution halfway to the first equivalence point. The same logic can be applied to absorption measurements. The protonated form of the indicator should be at 50 % of the total indicator concentration whenever the measured absorbance at 450 nm reaches 50 % of its maximum value.

This involves preparing a series of solutions having identical indicator concentrations but differing pHs, and measuring both the pH and absorbance of each solution. The experimental data would be graphed as Absorbance (y-axis) versus pH (x-axis). The change in the absorbance in going from the strongly acidic solutions to the strongly basic solutions corresponds to going from 100 % protonated to 0 % protonated. The absorbance of the solution with [In⁻]/[HIn] = 1 would have an absorbance equal to $0.5 \, \Delta A_{max}$ (see Figure 26.1). The absorbance of the unprotonated form need not be zero since the change in absorbance is used. A similar series of measurements could be performed at 550 nm, where the unprotonated species absorbs.

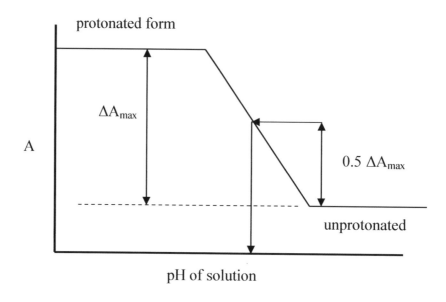

Figure 26.1. Determination of the equilibrium constant for an acid-base indicator from spectral absorption measurements.

SET UP THE PASCO SYSTEM FOR pH AND ABSORBANCE MEASUREMENTS

Enter the Pasco system **mouse clicking on the DataStudio icon** on the computer screen that comes up when the computer is first turned on. When the screen stating "How would you like to use DataStudio?" appears, **mouse click on Create Experiment**. Connect the Pas*Port* colorimeter and Pas*Port* pH/ORP/ISE Temperature sensor to the PowerLink unit. The system should recognize the accessories that are attached. To calibrate the pH electrode to the top toolbar and **mouse click on the Setup button**. The setup window should now open up on the screen. You will likely see arrows for both the colorimeter and pH/ORP/ISE electrode. **Click on the arrow to the left of the PH/ORP/ISE electrode** and proceed to calibrate the electrode using two pH buffers as you did in Experiments 15 – 18.

Next calibrate the colorimeter using deionzed water. Press the green oval button on the colorimeter and wait for the green light to go out. You will now want to modify the program to display Blue (468 nm) Absorbance and Green (565 nm) Absorbance. The program automatically records the measured intensity as % Transmittance, unless programmed otherwise. **Left mouse click on the Setup button** on the top toolbar. Maximize the Equipment window, and then mark the two absorbance readings that you want to appear. Once you have finished changing the display for the wavelengths and way that you want the recorded data displayed, **close the PasPort Sensors screen by mouse clicking on the lower X box screen**. Be careful not to close the DataStudio screen, which is the very top of the two X boxes.

EXPERIMENTAL pH AND ABSORBANCE MEASUREMENTS

It is now time to prepare the various solutions. The second color transition for bromothymol blue occurs in the pH = 6.2 to pH = 7.6 range. One will need to prepare several solutions (preferably buffered solutions) in this range. To prepare the buffers, one will need to find a suitable weak acid whose pKa falls somewhere between pH = 6.2 and pH = 7.6. Of the weak acids that you have studied thus far, phosphoric acid is probably the best choice. The second dissociation constant for phosphoric acid

$$H_2PO_4^-{}_{(aq)} + H_2O \quad \longleftrightarrow \quad H_3O^+{}_{(aq)} + HPO_4^{2-}{}_{(aq)}$$

$$K_{a2} = 10^{-7.2} = \frac{[H_3O^+][HPO_4^{2-}]}{[H_2PO_4^-]} \tag{26.2}$$

has a $pK_{a2} = 7.2$, which is in the desired pH range. An equimolar solution of KH_2PO_4 (or NaH_2PO_4) and Na_2HPO_4 (or K_2HPO_4) would have a pH of pH = 7.2. The ratio of

$[HPO_4^{2-}]/[H_2PO_4^-]$ could be adjusted from 1:10 to 10:1. This ratio would allow one to prepare several solutions of pH = 7.2 ± 1.0.

Prepare with pipette (or buret) the 7 solutions listed in Table 26.1. The solutions can be prepared by transferring the appropriate volumes of the various stock solutions into 100-ml clean, dry beakers. After the chemicals are thoroughly mixed, measure the absorbance of each solution at both Blue (468 nm) and Green (565 nm) wavelengths. Record the absorbance data in the appropriate space on the laboratory Data Sheet. The absorbances are measured first to avoid any possible contamination that might occur when the pH electrode is placed in the solutions. Most of the solutions are buffered, and the pH measurements should not be significantly affect if a small amount of solution is transferred from one beaker to the next.

Next measure the pH of each solution. In between measurements be sure to rinse the pH electrode by holding it over a beaker of waste solution, and squirting deionized water from a wash bottle onto the electrode. Gently blot the electrode with a paper towel before placing it in the next solution. Record the measured pH values on the Data Sheet.

After all of the experimental data has been recorded, plot the measured absorbance at 468 nm (y-axis) versus the pH of the solution (x-axis), and determine the value of pK_{HIn} from the graph at 0.5 ΔA_{max} (see Figure 26.1). Perform a similar analysis of the absorbance data measured at 565 nm.

Table 26.1. Preparation of Solutions for Bromothymol Blue Equilibrium Constant Measurement

Solution	Bromothymol blue 0.1 %	KH_2PO_4 0.1 Molar	Na_2HPO_4 0.1 Molar	H_2O
1	2.0 ml	10.0 ml	0.0 ml	38.0 ml
2	2.0 ml	10.0 ml	2.0 ml	36.0 ml
3	2.0 ml	20.0 ml	10.0 ml	18.0 ml
4	2.0 ml	10.0 ml	20.0 ml	18.0 ml
5	2.0 ml	2.0 ml	10.0 ml	36.0 ml
6	2.0 ml	2.0 ml	20.0 ml	26.0 ml
7	2.0 ml	0.0 ml	10.0 ml	38.0 ml

Compare your experimental value for the $pK_{HIn,2}$ for bromothymol blue to the published values that range from $pK_{HIn,2} = 7.1$ to $pK_{HIn,2} = 7.3$, as summarized in P. M. Sikia, M. Bora and R. K. Dutta, *J. Colloid Interface Sci.*, **285**, 382-387 (2005), and in J. S. Rhee, H.-M. Park, H. J. Kim, K.-B. Lee and P. K. Dasgupta, *Anal. Sci. Technol.*, **7**, 173-179 (1994).

DATA SHEET – EXPERIMENT 26

Pasco colorimeter

Name: _____

Measured Absorbance Data at 468 nm

1. Measured absorbance of solution 1: _____

2. Measured absorbance of solution 2: _____

3. Measured absorbance of solution 3: _____

4. Measured absorbance of solution 4: _____

5. Measured absorbance of solution 5: _____

6. Measured absorbance of solution 6: _____

7. Measured absorbance of solution 7: _____

Measured Absorbance Data at 565 nm

8. Measured absorbance of solution 1: _____

9. Measured absorbance of solution 2: _____

10. Measured absorbance of solution 3: _____

11. Measured absorbance of solution 4: _____

12. Measured absorbance of solution 5: _____

13. Measured absorbance of solution 6: _____

14. Measured absorbance of solution 7: _____

Measured pH Data

15. Measured pH of solution 1: _____

16. Measured pH of solution 2: _____

17. Measured pH of solution 3: _____

18. Measured pH of solution 4: _____

19. Measured pH of solution 5: _____

20. Measured pH of solution 6: _____

21. Measured pH of solution 7: _____

Equilibrium Constant Calculations

22. Equilibrium constant calculated from 468 nm data: _____

23. Equilibrium constant calculated from 565 nm data: _____

24. How does your experimental value for the $pK_{HIn,2}$ for bromothymol blue to the published values that range from $pK_{HIn,2} = 7.1$ to $pK_{HIn,2} = 7.3$?

EXPERIMENT 27: ENTHALPY OF VAPORIZATION DETERMINATION FROM VAPOR PRESSURE VERSUS TEMPERATURE MEASUREMENTS

INTRODUCTION

In today's laboratory experiment you will determine the enthalpy of vaporization of an alcohol by measuring how the alcohol's equilibrium vapor pressure varies with temperature. When a liquid is placed into a closed container the amount of liquid decreases. The decrease occurs because there is a net transfer of molecules from the liquid to the vapor phase. The evaporation process occurs at a constant rate at a given temperature. The reverse process is different, however. Initially, there are very few molecules in the vapor phase so the rate of return of gaseous molecules to the liquid state is less than the evaporation rate. Eventually, enough vapor molecules are present above the liquid so that the rate of condensation (vapor ---> liquid) exactly equals the rate of evaporation (liquid ---> vapor). At this point no further net change occurs in the amount of liquid or vapor because the two opposing processes balance each other. The system has now reached equilibrium. Equilibrium is a dynamic state. Molecules are still constantly escaping from and entering the liquid. A net change is not observed because the two rates are equal.

The pressure above the liquid at equilibrium is called the equilibrium vapor pressure, or more commonly, the vapor pressure of the liquid. The vapor pressures of liquids vary considerably. Liquids having high vapor pressures are said to be volatile, that is, they rapidly evaporate from an open container. Liquids with low vapor pressures are nonvolatile. Intermolecular forces in the liquid phase determine whether or not a given compound is volatile or nonvolatile. If the intermolecular forces between adjacent liquid molecules are large, then the vapor pressure will be low because molecules will need to have high energies in order to escape to the vapor phase. For example, water has a much lower molar mass than diethyl ether ($CH_3CH_2OCH_2CH_3$). The strong hydrogen-bonding forces that exist between water molecules in the liquid cause the vapor pressure of water (P_{water} = 23.76 torr at 25 °C) to be much lower than that of diethyl ether (P_{ether} = 536.7 torr at 25 °C). In general, substances with large molar masses have relatively low vapor pressures, mainly due to their large dispersion forces. The more electrons a substance possess, the more polarizable it is, and the greater the dispersion forces are. Polarity and hydrogen-bonding are important considerations as well.

Experimental measurements of the vapor pressure for a given liquid at several temperatures show that vapor pressure increases significantly with temperature. A plot of the natural logarithm of the equilibrium vapor pressure above a liquid, ln P , versus the reciprocal of the Kelvin temperature is linear

$$\ln P_{vap} = - (\Delta H_{vap}/R) (1/T) + intercept \qquad (27.1)$$

where ΔH_{vap} is the enthalpy of vaporization, and R is the universal gas constant. Equation 27.1 is referred to as the Clausius-Clapyeron equation, and it provides the mathematical basis for today's laboratory experiment. The vapor pressure of an alcohol will be measured at several different temperatures. From the slope of a ln P_{vap} versus 1/T plot the enthalpy of vaporization of the alcohol will be calculated.

CALIBRATE PASCO TEMPERATURE PROBE

Open up the Pasco system by mouse clicking on the Data Studio icon that appears on the computer screen. Now connect the Pasco temperature probe (pH/ORP/ISE Temperature Sensor PasPort) to the PowerLink unit. You will want the **Create Experiment** activity. The system should automatically recognize what probe(s) is (are) connected. After clicking on the Create Experiment activity, **Go to the top toolbar and click on the Setup button.** You will want to **click on the Calibrate button to the right of temperature.** Place the tip of the temperature in an ice-water both (which will serve as the 0 ºC reference data), **type in 0 for the Point 1 temperature**, wait a few minutes for thermal equilibrium to be reached, and then **mouse click on the Set button.** The lower reference temperature has now been set. For the higher temperature calibration point, use a heated water bath (above 80 ºC). You will need to use a wide temperature range for this experiment as the metal object for the heat capacity measurement will be placed in nearly boiling water. Place the Pasco temperature probe, along with a glass thermometer, into the hot water bath. Measure the temperature of the warm water with the glass thermometer. Now type this value into the box for the Point 2 temperature reading. Wait a few minutes for the temperature probe to achieve thermal equilibrium with the warm water. **Mouse click on the Set button. Now mouse click on OK button.** The Pasco temperature sensor is now calibrated.

VERIFICATION OF AMONTONS' LAW

In an earlier laboratory experiment (Experiment 7) the properties of gases were studied. Measurements were performed to verify Boyle's law, Charles' law and Avogadro's law. These three laws mathematically describe how the volume of a trapped gas sample varies with pressure, temperature and number of moles of gas, respectively. Now you will study how the pressure of a trapped gas varies with temperature.

Add 900 ml of tap water to a 1 liter beaker. Place the beaker on a hot plate. DO NOT TURN ON THE HOT PLATE YET. Next trap a sample of air by inserting the rubber-stopper assembly (a one-hole No. 4 or No. 5 rubber stopper with plastic tubing pushed through the hole. The tubing should fit tightly into the hole to make an airtight seal) into the mouth of a clean, dry 125-ml Erlenmeyer flask. **Important**: Twist the stopper into the neck to ensure a tight fit. Connect the rubber tubing to the Absolute Pressure Sensor Pas*Port*. Connect the Absolute Pressure Sensor Pas*Port* to the PowerLink unit. Now slide the hot plate and beaker close to the PowerLink unit. Put the temperature probe into the water bath. Be sure that the probe is held up so that it does not touch the bottom of the beaker. Using a pair of crucible tongs hold the capped Erlenmeyer flask in the water bath. As much of the flask as possible, except for the very top and rubber stopper, needs to be in the water until the last measurement is made to ensure that the temperature of vapor is the same as the water bath. (See Figure 27.1 for the apparatus.) Be sure that the plastic tubing and temperature probe wire does not touch the top hot plate surface. Turn the hot plate to high and **mouse click on the start button** on the top tool bar. Record the initial temperature and pressure on the laboratory data sheet in the section labeled "Verification of Amonton's Law." Continue recording the pressure at five-degree intervals from 25 °C to 65 °C. Be sure to enter the values on the Data Sheet. Once the last pressure is recorded, **mouse click on the Stop button**, and turn off the hot plate.

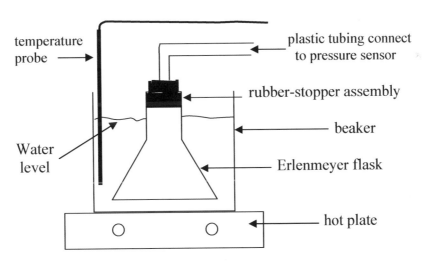

Figure 27.1. Set up of apparatus for the enthalpy of vaporization determination.

Carefully remove the Erlenmeyer flask from the hot water bath and disconnect the plastic tubing from the Absolute Pressure Sensor Pas*Port*. Slide the hot plate and beaker out of the way. BE CAREFUL NOT TO TOUCH ANY OF THE HOT SURFACES.

You are now ready to analyze the Pressure versus Temperature data. On the left hand side of the monitor screen, **mouse click on the Graph line** in the Display window.

309

A new window should open up in the center of the monitor screen. **Mouse click on the Temperature – Run #1 line.** A graph of Temperature versus Time should now be displayed. The data that is plotted on the y-axis can be changed by **moving the mouse cursor over the label Temperature** – a box with vertical lines (≡) should appear. **Left-click on the mouse** – a window with several options should appear. **Scroll down to Pressure and mouse click.** The y-axis should now correspond to pressure. The x-axis can be changed in similar fashion **by moving the mouse cursor to the label Time (s).** A box with several vertical lines (≡) should appear. **Scroll down to Temperature and mouse click.** The x-axis should now be changed to temperature. The linear regression is performed on the Pressure versus Temperature data by **mouse clicking on the Fit button at the top of the Graph tool bar, and then scrolling down to Linear Fit.** Mouse click. Record the slope, intercept and correlation coefficient in the appropriate place on the laboratory data sheet. You will need this information to calculate the Centigrade temperature corresponding to absolute zero. Substitute the numerical values of the slope and intercept into Equation 27.2

$$\text{Pressure} = \text{slope} \times \text{Temperature (in °C)} + \text{intercept} \qquad (27.2)$$

Set the pressure equal to zero, and solve for the Centigrade temperature that corresponds to absolute zero, *i.e.*, 0 Kelvin. William Thompson (also known as Lord Kelvin) is recognized for having proposed an absolute-temperature scale in 1848. Several years earlier, in 1779 Joseph Lambert defined absolute zero as the temperature at which the pressure of a gas becomes zero when a plot of gas pressure versus temperature is extrapolated to $P_{gas} = 0$. (See L. H. Adcock, *J. Chem. Educ.*, **75**, 1567-1568 (1998) for additional explanation.) How would the value of absolute zero defined in this manner theoretically compare to the value based on Charles' law (see Experiment 7)?

ENTHALPY OF VAPORIZATION MEASUREMENTS

The experiment will now be repeated with alcohol inside the Erlenmeyer flask. First empty the hot water from the 1 liter beaker, and refill with 900 ml of tap water. The water that was in the beaker from the first series of measurements is too warm for the lower temperature-pressure measurements. Place the beaker on a hot plate. DO NOT TURN ON THE HOT PLATE YET. Obtain a new clean, dry 125-ml Erlenmeyer flask. Trap an air sample by inserting the rubber-stopper assembly into the mouth of the flask. Connect the end of the plastic tubing to the Pressure Sensor Pas*Port*. Sit the beaker and temperature probe on the laboratory bench to equilibrate. After 10 minutes have passed, mouse click on the Start button on the top toolbar. Record the temperature and pressure on the Laboratory Data Sheet under the section heading labeled "Enthalpy of Vaporization Measurements (First Alcohol)". The pressure recorded on line number 11 of the Data Sheet represents the pressure of the air that will be inside the flask at the start of the enthalpy of vaporization measurements. You will need to recalculate the air pressure for each of the nine temperatures studied, using

$$P_{air} \text{ (at T)} = P_{air} \text{ (at } T_{initial}) \, (T / T_{initial}) \tag{27.3}$$

where $T_{initial}$ is the value of the temperature (in Kelvin) recorded on line number 10 of the Data Sheet. Remember from the first set of pressure measurements that the pressure of air does increase linearly with temperature. Equation 27.3 represents the Amontons' law mathematical relationship between pressure and temperature. (Pressure is directly proportional to the absolute temperature.)

Remove the rubber-stopper assembly from the Erlenmeyer flask, quickly add 10 ml of alcohol (with graduated cylinder) to be studied, and immediately replace the rubber-stopper assembly on the flask. You need to compete this step as fast as possible so that very little alcohol (hopefully none) evaporates inside the flask while the rubber stopper is off. The calculations assume that the pressure of air inside the flask is the same as the value initially measured. In reality, the pressure inside the flask at this time is that of the atmosphere, $P_{inside\ flask} = P_{air} + P_{alcohol}$, and if no alcohol evaporated ($P_{alcohol} = 0$), then P_{air} inside the flask should be the numerical value that you recorded on line number 11 of the Data Sheet.

Now slide the hot plate and beaker close to the PowerLink unit. Put the temperature probe into the water bath. Be sure that the probe is held up so that it does not touch the bottom of the beaker. Using a pair of crucible tongs hold the capped Erlenmeyer flask in the water bath. As much of the flask as possible, except for the very top and rubber stopper, needs to be in the water until the last measurement is made to ensure that the temperature of vapor is the same as the water bath. Be sure that the plastic tubing and temperature probe wire does not touch the top hot plate surface. Turn the hot plate to high and **mouse click on the start button** on the top tool bar. Record the pressure at five-degree intervals from 25 °C to 65 °C. Once the last pressure is recorded, **mouse click on the Stop button**, and turn off the hot plate. Carefully remove the Erlenmeyer flask from the hot water bath and disconnect the plastic tubing from the Absolute Pressure Sensor Pas*Port*. Slide the hot plate and beaker out of the way. BE CAREFUL NOT TO TOUCH ANY OF THE HOT SURFACES.

Ask your TA if there is enough time for you to study a second alcohol, and still be able to perform the linear regression. If there is sufficient time, repeat the enthalpy of vaporization measurements with a different alcohol. The three alcohols that can be conveniently studied are: methanol ($\Delta H_{vap} = 36.6$ kJ/mole); ethanol ($\Delta H_{vap} = 41.2$ kJ/mole); and 2-propanol ($\Delta H_{vap} = 43.9$ kJ/mole). A literature value for the enthalpy of vaporization of each alcohol is given in parenthesis.

Once all of the enthalpy of vaporization measurements is complete, you will need to determine the slope of the ln Pressure of alcohol versus 1/Temperature plot(s) through linear regression analysis. The slope is needed to calculate the enthalpy of vaporization. The pressures that were measured above the alcohol sample corresponded to the sum of the pressure of alcohol plus the pressure of the trapped air. The pressure of the trapped air needs to be corrected for the change in temperature using Eqn. 27.3. The pressure of the alcohol at each temperature, $P_{alcohol}$, is

$$P_{alcohol} = P_{measured} - P_{air} \text{ (at T)} \tag{27.4}$$

calculated by subtracting the value of the corrected air pressure from the measured total pressure. There is space on the Data Sheet (line number 13 for first alcohol and line number 24 for the second alcohol) for recording these values. Also calculate the ln $P_{alcohol}$. Once all of the values are calculated, you can perform the linear regression using the Pasco software, or using an Excel spreadsheet. If you decide to use the Pasco system, you can either exit the current activity, or you can do the analysis within the current activity (Experiment button on the top toolbar – New Empty Data Table – described in greater detail in Appendix A). To do a linear least squares regression analysis, go up to the **File button on the top toolbar, click**, and **scroll down to New Activity and mouse click**. There should be no reason to the store the experimental temperature and pressure data as it should be recorded on the Data Sheet. **Click on the Enter Data picture** on the Welcome to DataStudio screen. Enter the 1/T (x-value) and ln $P_{alcohol}$ (y-value) for the first alcohol. After all of the data has been entered, the linear fit is obtained **by mouse clicking on the Fit button** at the top of the graph toolbar, and **scrolling down to Linear fit**. Record the slope of the straight line and the correlation coefficient on your laboratory Data Sheet. The enthalpy of vaporization is calculated by multiplying the slope of the line by –8.314 J/(mole K). Divide by 1000 to convert the value to kJ/mole.

The value that you obtain for the enthalpy of vaporization will likely greater than the literature value. This is to be expected because there are dissolved gases (oxygen, nitrogen, *etc.*) in the alcohols. As the liquid alcohol samples are heated, the solubility of the dissolved gases decrease. The undissolved oxygen, nitrogen, *etc.* gases contribute then to the measured pressure. The value that is being calculated for $P_{alcohol}$ is then too large, as is the slope in the ln $P_{alcohol}$ versus 1/T plot. There is no convenient way to correct for the effect of dissolved gases. One could boil the liquid alcohol(s) to "outgas" the sample(s) prior to the pressure measurements, but this will be fairly time-consuming as one would have to wait for the alcohol(s) to cool back to room temperature. The alcohol would also have to cool in a sealed container to prevent oxygen, nitrogen, *etc.* gas from dissolving back in the sample as it cools.

DATA SHEET – EXPERIMENT 27

Name: _____

Amontons' Law Measurement:

1. Initial temperature, °C: _____

2. Initial pressure, kPa: _____

3. Pressure of air at specified temperatures:

Temperature (°C)	Pressure (kPa)
_____	_____
25.0	_____
30.0	_____
35.0	_____
40.0	_____
45.0	_____
50.0	_____
55.0	_____
60.0	_____
65.0	_____

4. Sketch of the Pressure versus Temperature plot:

Pressure
(kPa)

Temperature (°C)

5. Slope of pressure versus temperature plot: _____

6. Intercept of pressure versus temperature plot: _____

7. Correlation coefficient of pressure versus temperature plot: _____

8. Calculated value for absolute zero, °C: _____

313

Name: _____

Enthalpy of Vaporization Measurement (First Alcohol):

9. Name of alcohol: _____

10. Initial temperature, °C: _____

11. Initial pressure, kPa: _____

12. Total pressure above alcohol at specified temperatures:

Temp. (°C)	Pressure (kPa)
25.0	_____
30.0	_____
35.0	_____
40.0	_____
45.0	_____
50.0	_____
55.0	_____
60.0	_____
65.0	_____

13. Calculated values for ln Pressure versus 1/T graph:

1/Temp. (K)	P_{air} (kPa)	$P_{alcohol}$ (kPa)	ln $P_{alcohol}$
0.0033540	_____	_____	_____
0.0032987	_____	_____	_____
0.0032452	_____	_____	_____
0.0031934	_____	_____	_____
0.0031432	_____	_____	_____
0.0030945	_____	_____	_____
0.0030474	_____	_____	_____
0.0030017	_____	_____	_____
0.0029573	_____	_____	_____

14. Slope of pressure versus temperature plot: _____

15. Intercept of pressure versus temperature plot: _____

16. Correlation coefficient of pressure versus temperature plot: _____

17. Calculated value of the enthalpy of vaporization, kJ/mole: _____

18. Literature value of the enthalpy of vaporization, kJ/mole: _____

19. Percent deviation: _____

Enthalpy of Vaporization Measurement (Second Alcohol):

20. Name of alcohol: _____

21. Initial temperature, °C: _____

22. Initial pressure, kPa: _____

23. Total pressure above alcohol at specified temperatures:

Temp. (°C)	Pressure (kPa)
25.0	_____
30.0	_____
35.0	_____
40.0	_____
45.0	_____
50.0	_____
55.0	_____
60.0	_____
65.0	_____

24. Calculated values for ln Pressure versus 1/T graph:

1/Temp. (K)	P_{air} (kPa)	$P_{alcohol}$ (kPa)	ln $P_{alcohol}$
0.0033540	_____	_____	_____
0.0032987	_____	_____	_____
0.0032452	_____	_____	_____
0.0031934	_____	_____	_____
0.0031432	_____	_____	_____
0.0030945	_____	_____	_____
0.0030474	_____	_____	_____
0.0030017	_____	_____	_____
0.0029573	_____	_____	_____

25. Slope of pressure versus temperature plot: _____

26. Intercept of pressure versus temperature plot: _____

27. Correlation coefficient of pressure versus temperature plot: _____

28. Calculated value of the enthalpy of vaporization, kJ/mole: _____

29. Literature value of the enthalpy of vaporization, kJ/mole: _____

30. Percent deviation: _____

Name: _____

Class Data – Enthalpy of vaporization for first alcohol:

_____	_____	_____	_____
_____	_____	_____	_____
_____	_____	_____	_____
_____	_____	_____	_____
_____	_____	_____	_____
_____	_____	_____	_____

Class Mean for vaporization enthalpy of first alcohol: _____

Class Standard deviation for vaporization enthalpy of first alcohol: _____

Class Data – Enthalpy of vaporization for second alcohol:

_____	_____	_____	_____
_____	_____	_____	_____
_____	_____	_____	_____
_____	_____	_____	_____
_____	_____	_____	_____
_____	_____	_____	_____

Class Mean for vaporization enthalpy of second alcohol: _____

Class Standard deviation for vaporization enthalpy of second alcohol: _____

EXPERIMENT 28: PERCENTAGE OF OXYGEN IN AIR BASED ON DALTON'S LAW OF PARTIAL PRESSURES

INTRODUCTION

In Experiment 7 you verified Boyle's law and Charles' law by measuring how the volume of a trapped gas varies with changes in the pressure and absolute temperature, respectively. Volume was found to be inversely proportional to pressure (Boyle's law), and to be directly proportional to temperature expressed in Kelvin (Charles' law). Experimental volume versus pressure and volume versus temperature observations, when combined the measured volume versus moles of trapped gas data, led to the ideal gas law. The ideal gas law states that the pressure, P, times volume, V, product equals

$$P \times V = n \times R \times T \qquad (28.1)$$

where n denotes the number of moles of gas, T refers to the absolute temperature (in Kelvin), and R is the universal gas constant (R = 0.08206 liter-atm/(mole K)).

John Dalton, while studying the properties of air in the early 1800's, observed that the total pressure of a mixture of gases equals the sum of the pressures that each gas would exert if it were present alone. For example, in a gas mixture containing only nitrogen and oxygen gas, the total pressure, P_{total}, is

$$P_{total} = P_{nitrogen\ gas} + P_{oxygen\ gas} \qquad (28.2)$$

Equation 28.2 is the mathematical statement of Dalton's law of partial pressures, which forms the basis of today's laboratory experiment.

The experiment will involve trapping a sample of air in the presence of an activated iron metal surface. The oxygen gas in the air will react with the iron metal

$$2\ Fe_{(solid)} + 1.5\ O_{2(gas)} \ \ \text{---->}\ Fe_2O_{3(solid)}$$

to form an iron oxide. The pressure of the trapped gas decreases as oxygen reacts with the metal surface. Once all of the oxygen is removed, the pressure should remain constant. By measuring the initial and final pressures, P and P , one should be able to calculate

$$P_{oxygen\ gas} = P_{initial} - P_{final} \qquad (28.3)$$

the partial pressure of oxygen gas in air.

Dry air is predominately a mixture of nitrogen and oxygen gases. The other gases present (argon, carbon dioxide, neon, helium, carbon monoxide, methane, *etc.*) total less than one percent by mole fraction. Assuming that each gas behaves ideally, one can express the partial pressure of each gas

$$P_{nitrogen\ gas} = n_{nitrogen\ gas} \times R \times T / V \qquad (28.4)$$

$$P_{oxygen\ gas} = n_{oxygen\ gas} \times R \times T / V \qquad (28.5)$$

$$P_{argon\ gas} = n_{argon\ gas} \times R \times T / V \qquad (28.6)$$

$$\bullet \bullet \bullet \bullet, \text{ etc.}$$

by Eqn. 28.1. All gases present in the air sample that you trapped are at ambient room temperature, and all gases occupy the same volume. Substitution of the individual partial pressure expressions into

$$P_{total} = P_{nitrogen\ gas} + P_{oxygen\ gas} + P_{argon\ gas} + \ldots\ldots \qquad (28.7)$$

gives

$$P_{total} = (n_{nitrogen\ gas} + n_{oxygen\ gas} + n_{argon\ gas} + \ldots..) \times R \times T / V \qquad (28.8)$$

$$P_{total} = n_{total} \times R \times T / V \qquad (28.9)$$

For a mixture of ideal gases it is the total of moles of particles that is important, not the identity nor composition of the involved gas particles. The mole fraction of oxygen gas in air, $x_{oxygen\ gas}$, can be calculated by

$$x_{oxygen\ gas} = n_{oxygen\ gas} / n_{total} \qquad (28.10)$$

$$x_{oxygen\ gas} = P_{oxygen\ gas} / P_{total} \qquad (28.11)$$

$$x_{oxygen\ gas} = (P_{initial} - P_{final}) / P_{initial} \qquad (28.12)$$

VAPOR PRESSURE MEASUREMENTS

Pressure will be measured using the Pasco pressure transducer. Enter the Pasco system by **double left-hand mouse clicking on the DataStudio icon** on the computer screen. When the screening saying "How would you like to use DataStudio" appears, **mouse click on Create Experiment**. Connect the Pasco Absolute Temperature-Pressure Sensor to the PowerLink unit. The system should automatically recognize what accessories are attached. If for some reason the system does not recognize that the Absolute Temperature-Pressure Sensor is attached, disconnect and reconnect the sensor to the PowerLink unit. If this does not work, exit the system and re-enter by mouse clicking on the DataStudio icon. There is no calibration of the pressure sensor per se. The measured pressure can be recorded in kPa as the computations involve a pressure ratio, and the units will cancel.

Measure the initial pressure of air by **mouse clicking on the Start button** at the top. Once a steady value is obtained, record the initial pressure on the Laboratory Data Sheet and mouse click on the Stop button. You will now need to weigh out approximately 1.0 gram of steel wool (coarse steel wool works fine). The mass does not enter into calculation; however, one does need to use a sufficient amount of steel wool so that the oxygen is removed in timely fashion. Soak the steel wool in 50 ml of acetone for 30 seconds, remove and thoroughly dry with a paper towel. The acetone soak should remove any organic material that might be on the surface of the steel wool. Next swirl the steel wool in 50 ml of vinegar for about 30 seconds. Remove the steel wool from the vinegar and thoroughly dry with a paper towel. Repeat the washing with a dilute vinegar solution (10 ml of vinegar diluted to 100 ml with deionized water.) Thoroughly dry with a paper towel. (NOTE – If the steel wool is not sufficiently dried of the acid, hydrogen gas may be generated from the reaction of the iron in the steel wool with the acid. There is sufficient acid still left on the steel wool after "thoroughly drying" to catalyze the oxygen gas reaction.) After drying, immediately slide the steel wool into the test tube (125 mm). The steel wool should remain as spread out as possible in order to maximize the reaction surface area. A large exposed surface area is desired in order to reduce the reaction time. Once the steel wool has been slid into the test tube, place the small rubber stopper and Tygon tubing assembly into the mouth of the test tube, and then connect the tubing to the Pressure Sensor PasPort. Doing the connections in this order should reduce any extra pressure caused when the tubing is connected to the pressure sensor. (See Figure 28.1 for a diagram of the assembled apparatus.)

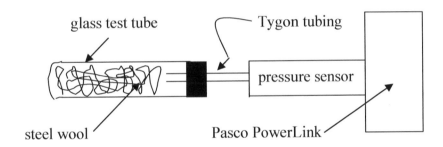

Figure 28.1. Assembled apparatus for determining the percentage of
Oygen gas in air. The glass test tube is connected to the Pressure
Sensor PasPort device by a short piece of Tygon tubing.

Mouse click on the Start button at the top to start the pressure measurements. The pressure should decrease fairly rapidly at first, and then gradually tapper off. RECORD THE PRESSURE FOR A MINIMUM OF 45 MINUTES. You will need to make an educated guess when the reaction is finished as evidenced by the attainment of a constant pressure. To view the pressure graphically, **double left-hand mouse click on the Graph line in the Display section** on the left-hand side of the computer screen. **Scroll down to Run #2 and mouse click**. (Run #1 should be the initial pressure measurement.) You should now be able to see a graph of the measured pressure as a function of time, much like the graph depicted in Figure 28.2. Once the pressure has

319

reached a steady value, **mouse click on the Stop button** at the top. The final pressure can be read off the graph by **mouse clicking on the Smart Tool button on the Graph Toolbar** (sixth box from the left-hand side). An "+- grid" appears, and one can move the grid to specific parts of the pressure versus time curve by positioning the mouse cursor in the center of the grid, holding down the mouse button, and then dragging the cursor to the final set of experimental data points. Record the final pressure on the Data Sheet, and calculate the mole fraction percentage of oxygen in air by substituting the initial and final pressures into Eqn. 28.12. Give the value of the mole fraction percentage of oxygen in air to the TA who will then compile the values for the entire class. The data for the entire class will be posted on the bulletin board outside the laboratory room sometime during the next day. Go to the bulletin board outside the laboratory room and copy down the class results on your Laboratory Data Sheet in the space provided. Calculate the class mean (average) and standard deviation for the percentage of oxygen gas in air. Use the Q-test to reject any outlier values before computing the average and standard deviation.

The mole fraction of oxygen gas in dry air is reported to be $x = 0.2095$ (or so). How does your experimental value compare to the literature value?

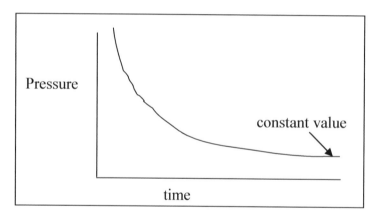

Figure 28.2. Graph of the measured total pressure versus time graph. The pressure measurements can be stopped when the total pressure reaches a constant value. NOTE - RECORD THE PRESSURE FOR A MINIMUM OF 45 MINUTES

DATA SHEET – EXPERIMENT 28

P_{He} = 2 atm

P_{Ar} = 1 atm

P_{Ne} = 3 atm P_{total} = 6 atm

Name: _____

Percentage of oxygen gas in air measurements:

1. Initial air pressure, kPa: _____
2. Final air pressure, kPa: _____
3. Partial pressure of oxygen gas, kPa: _____
4. Mole fraction of oxygen gas in air: _____
5. Literature value for the mole fraction of oxygen gas in air: _____
6. Percent deviation: _____
7. Explain in a brief paragraph why your experimental value might differ
    from the literature value.

Class Data – Mole fraction of oxygen gas in air:

_____	_____	_____	_____
_____	_____	_____	_____
_____	_____	_____	_____
_____	_____	_____	_____
_____	_____	_____	_____
_____	_____	_____	_____

Could any of the experimental values be rejected by the Q-test, if so, which value(s):

_____   _____   _____   _____

Class Mean for mole fraction of oxygen gas in air: _____

Class Standard deviation for mole fraction of oxygen gas in air: _____

APPENDIX A: PASCO SYSTEM CALIBRATION AND OPERATING INSTRUCTIONS

ABSORBANCE MEASUREMENTS

Enter the Pasco system by **clicking on the DataStudio icon** on the computer screen. When asked by the system "How would you like to use DataStudio?" **click on Create Experiment**. Connect the Pas*Port* Colorimeter to the PowerLink unit. To zero the instrument fill the glass sample cell with distilled water (or appropriate solvent blank). Open the top of the sample compartment on the colorimeter unit, insert the sample tube of distilled water, and close the lid completely. You do not want any stray radiation from the room reaching the detector. On the Pas*Port* Colorimeter probe, between the sample holder lid and where the unit is connected into the PowerLink box, you will see a green oval button. **Press the oval green calibrate button**. A green light should illuminate on the oval button to indicate that the calibration is in progress. Wait for the light to turn off and then remove the sample. [Note- a blinking red light on the oval calibration button means either: (a) the dark count is too high; (b) stray light is entering the sample compartment; or (c) the sensor measurement is out of range. The light turns off when the reading is within the normal range. To verify that the instrument is indeed calibrated, reinsert the sample of distilled water, close the lid, and click the Start button on the top tool bar. All color readings should be approximately 100 % transmittance (0 absorbance if the system is set up to display absorbance.)

The Pasco software automatically records the measured intensity data as % Transmittance (if you need to make a conversion to absorbance, the formula is Absorbance = 2 – log % Transmittance). To modify the program to display the measured data as absorbances, left mouse click on the Setup button on the top toolbar. Off to the right a new screen titled "Equipment Setup" should appear. Maximize the "Equipment Setup" screen by mouse clicking on the ☐ box in the upper right-hand corner of the new screen. When the screen is maximized you should see a series of eight boxes that look as follows:

        ☐ Red (660 nm)   Transmittance %

        ☐ Green (565 nm)   Transmittance %

        ☐ Blue (468 nm)   Transmittance %

        ☐ Orange (610 nm) Transmittance %

        ☐ Red (660 nm)   Absorbance

        ☐ Green (565 nm)   Absorbance

        ☐ Blue (468 nm)   Absorbance

        ☐ Orange (610 nm) Absorbance

An "X" in the box will show what the instrument is currently set up to display. To change the display format, point the mouse cursor in the box and click. Off to the upper left-hand side of the computer screen, under the box titled DataScreen, you should see new types of wavelength readings (color, transmittance or color, absorbance) appearing and disappearing as they changed. For most of applications you will want to display the absorbance as this is the quantity that is related to concentration through Beer's law. Once you have finished changing the display for the wavelengths and way that you want the recorded data displayed, close the PasPort Sensors screen by mouse clicking on the lower of the two X boxes near the upper right-hand corner of the computer screen. **Be careful not to close the DataStudio screen, which is the very top of the two X boxes.**

Now that the system is set up to record and display the intensity data that you want to measure, you need to tell the system whether you want the data to be displayed as in a time versus absorbance table format or as a plot of absorbance (y-axis) versus time (x-axis). If you want the data to be displayed in table format, go to the Display section of the left-hand side of the computer screen. **Double left-hand mouse click on the Table line.** A new window will open. Scroll down to the wavelength (color) and type of measurement (Transmittance % or absorbance), and **mouse click OK**. The measured values will be displayed as ordered (x,y) pairs in a time-absorbance table.

If you want the measured intensity data to be displayed in a graphical format of time (x-axis) versus Transmittance % (or Absorbance) (y-axis) double left-hand mouse **click on the Graph button** in the Display section of the left-hand side of the computer screen. Scroll to the wavelength (color) and type of measurement (Transmittance % or absorbance) you want displayed and **mouse click OK**. The system should now be set up to take make experimental measurements.

AVERAGE AND STANDARD DEVIATION COMPUTATIONS

The Pasco system has the capability of performing statistical calculations. To calculate and average and standard deviation using the built-in software, double on Data Studio. You have now opened the system. **Double click on the activity titled Enter Data.** What should now appear on the screen is a table on the left-hand side, and a graph on the right-hand side. Enter the values to be averaged in the x-column of the table. Don't worry about what is happening with the y-column. Once all of the values are entered, **click on the Σ key on the Table toolbar**. The mean should now be displayed at the bottom of the column. If you want to calculate the standard deviation, go back up to the top of the table, and **click on the ▼ arrow by the Σ key**, scroll down to standard deviation, and hit enter. The standard deviation is now at the bottom the page, along with the average. When you get ready to enter the next set of data, you can either the enter the next set right over the previous data set (be sure to delete out any extra data points), or you can clear the numerical values in the table by **mouse clicking on New Data** on the top tool bar.

CHANGE NAME OF A GRAPH OR TABLE

To change the name of a graph or table, go to the Displays box and click twice (slowly) on the graph (or table) name that you would like to change. A prompt will appear, that will allow you to change name. Press "ENTER" when done.

CLEARING RECORDED DATA RUNS FROM TABLES

As you accumulate experimental data from pH measurements, absorption measurements, *etc.* from multiple runs it will be necessary to periodically clear the entries from the screen. The Table screen is only so large, and after awhile the columns start to overlap. To clear data in between runs, **mouse click on the Data ▼button** on the Table toolbar. The data that is being displayed in the table will have a √ check-mark to the left of the Run number. Scroll down to the data that you no longer want to display, **mouse click on the run number**, and the √ check-mark should disappear. The runs that you have unmarked should no longer be displayed. The data is still stored, however, and can be easily retrieved by **clicking on the Data ▼button** on the table toolbar. Mark the data that you want redisplayed by placing the cursor of the mouse pointer on the run number and **click OK**. The numbers will be redisplayed in the table.

COOLING CURVE MEASUREMENTS

Open up the Pasco system by mouse clicking on the Data Studio icon that appears on the computer screen. Now connect the Pasco temperature probe (pH/ORP/ISE Temperature Sensor Pas*Port*) to the PowerLink unit. You will want the **Create Experiment** activity. The system should automatically recognize what probe(s) is (are) connected. After clicking on the Create Experiment activity, **Go to the top toolbar and click on the Setup button.** You will want to **click on the Calibrate button to the right of temperature.** Place the tip of the temperature in an ice-water both (which will serve as the 0 °C reference data), **type in 0 for the Point 1 temperature**, wait a few minutes for thermal equilibrium to be reached, and then **mouse click on the Set button**. The lower reference temperature has now been set. For the higher temperature calibration point, use a heated water bath (above 50 °C). Place the Pasco temperature probe, along with a glass thermometer, into the warm water bath. Measure the temperature of the warm water with the glass thermometer. Now type this value into the box for the Point 2 temperature reading. Wait a few minutes for the temperature probe to achieve thermal equilibrium with the warm water. **Mouse click on the Set button. Now mouse click on OK button.** The Pasco temperature sensor is now calibrated.

Now that the system is set up to record and display the temperature data that you want to measure, you need to tell the system whether you want the data to be displayed as in a time versus temperature table format or as a plot of temperature (y-axis) versus time (x-axis). For cooling curve measurements you will want the data displayed in graphical format. Go to the Display section of the left-hand side of the computer screen. **Double left-hand mouse click on the Graph line.** A new window will open. Scroll down to the type of measurement (Temperature), and **mouse click OK**. The measured values will be displayed in graphical format.

DROP COUNTER CALIBRATION

Attach the two plastic stopcocks one above the other onto the plastic syringe buret. The top stopcock will be used to adjust the drip rate. The bottom stopcock will be used to turn the titrant flow on (full vertical position) or off (full horizontal position). Once the drip rate is set **DO NOT TURN THE TOP STOPCOCK AGAIN**. This would change the drip rate. Fill the syringe with the deionized water. Open the bottom stopcock to the full vertical position. Now slowly turn the top stopcock until you get a steady drop rate of 3 to 4 drops per second. Once this is achieved, turn the bottom stopcock to the horizontal position. **DO NOT TURN THE TOP STOPCOCK AGAIN UNTIL YOU ARE READY TO ADJUST THE DRIP RATE.**

Now you need to tell the system how large the falling drops are. Position the syringe buret over the rectangular opening in the Drop Counter. Leave room under the drop counter to place the titration beaker and a graduated cylinder. Place a graduated cylinder under the Drop Counter, aligned with the buret so that the dripping titrant will fall into the graduated cyclinder. The cylinder will be used to measure the volume of the titrant that is delivered from the buret. **Mouse click on the Setup button** on the top toolbar. The setup window will now reappear on the screen. **Mouse click on the ▼ or ▶ symbol** to the left Drop Counter. The calibrate screen should now appear on the screen. Open the bottom stopcock to allow the liquid drops to fall directly into the graduated cylinder. You do not want the liquid to drops to hit the Drop Counter or to miss the beaker. If for some reason you need to turn of the buret before the calibration is finished, or if the drops miss the graduated cylinder, turn the bottom stopcock off. You will need to empty any liquid from the buret, close the calibration screen, and then re-enter the Drop Counter calibration screen by mouse clicking on the Setup button on the top tool bar. Once the liquid in the graduated cylinder reaches the 10-ml mark, turn of the bottom stopcock. You now need to tell the system how much liquid actually flowed between the Drop Counter cells. Since 10-ml of liquid was collected in the graduated cylinder, **type 10.00 in the box on the calibration screen. Mouse click on the Set button on the calibration screen.** The number that is above the box should change to 10.00 (e.g., the number that you typed). When this has been done, **mouse click OK**.

The green light on the Drop Counter should be flashing the entire time the drops are falling. If the green light stays on (does not flash), this indicates that some liquid has splashed onto the lens. Should this happen gently position a paper towel into the rectangular box, and move along the inside walls to remove the splashed water.

FIRST DERIVATIVE OF pH TITRATION CURVE

To access the built-in software you need to have the pH titration curve displayed on the computer screen. **Mouse click on the Calculator button** (first button to the right of Fit ▼) on the Graph screen. You now need to define the mathematical function you want to calculate and display. In the definition box type $y =$. (Remove anything else that might be in the box.) **Mouse click on the Special ▼ button**. A list of options will appear. **Scroll down to derivative (2,x) and mouse click**. In the definition box there should now be y = derivative(2,x). Now you need to tell the software what data to take the first derivative of. **Click on the ▼ to the left of x = pH versus Fluid Volume** in the Variables box. A new screen should appear. **Scroll down to Data Measured and mouse click**. A new screen should appear. **Scroll down to pH versus Fluid Volume and mouse click OK**. Now **mouse click on the √ Accept button**. Close the calculator screen and the first derivative curve should appear on the graph.

LINEAR LEAST SQUARES ANALYSIS

A linear least squares analysis using the Pasco system is performed as follows: Enter the Pasco system by **mouse clicking on the DataStudio icon** that is displayed on the computer screen when the computer is first turned on. You should now see what is called the Welcome to DataStudio screen. **Right mouse click on the Enter Data picture**. On the left-hand side of the screen there should be an x-y data table, and on the right-hand side of the screen there should be an x-y graph. Enter the x-value of the first data point in the first line of the x-column, hit enter, type the y-value of the first data point, enter, type the x-value of the second data point, enter, ……., until all of the data points have been entered. Examine the table carefully for typographical errors. To perform the least squares analysis, **click on the Fit button** at the top of the Graph side toolbar, **scroll down to Linear Fit**, and left click on the mouse. The slope and the intercept of the line, as well as the correlation coefficient, are now displayed inside the box on the computer screen.

To perform the next linear least squares analysis, go up to the file button at the top left-hand corner of the toolbar. **Mouse click on the File Button, scroll down to New Activity, and mouse click**. When the system asks "Should DataStudio save this activity? Click no. You should now be back to the Welcome to DataStudio screen where you began the first least squares analysis calculation.

MOVING TABLES AND GRAPHS TO OTHER LOCATIONS

Tables and graphs can be moved to other locations of the computer screen by placing the mouse cursor on the blue bar at the top of the table or graph, holding down the mouse button, and then dragging the table or graph to the new location.

pH MEASUREMENTS

Enter the Pasco system by **mouse clicking on the DataStudio icon** on the computer screen that comes up when the computer is first turned on. When the screen saying "How would you like to use DataStudio" comes up, **mouse click on Create Experiment**. Connect the pH probe to the PowerLink unit. The system should recognize what accessories have been attached. Go to the top tool bar and **mouse click on the Setup button**. The setup window will now open up on the screen. The set up window will be for calibrating either the pH probe or the drop counter. One can go back and forth between the two calibrations by clicking on the ▼ or ▶ symbol to the left of the accessory name. On the screen for the pH/ISE/Temperature sensor, there will be three options displayed on the screen. You will want to **click on the Calibrate button directly across from the pH box**. The pH calibration is a two-point calibration. Place the pH electrode in the first pH standard/buffered solution, **type in the pH of the buffer in the box for Point 1**, wait a few minutes for the pH electrode to reach equilibrium with the solution, and then **mouse click on the Set button**. The first pH point is now set. Take the pH electrode out of the standard solution, hold the electrode over an empty beaker, rinse the electrode with deionized water from your wash bottle, and carefully blot dry with a paper towel, before placing in the second pH standard/buffer solution. Although the pH of a buffered solution is not suppose to change upon addition of a small amount of acid or base, it is good practice to rinse the pH electrode with deionized before placing it into the next solution. The pH of Point 2 is set by placing the pH electrode into the second standard/buffered solution, **typing in the pH in the box for Point 2**, waiting a few minutes for the electrode to reach equilibrium with the buffered solution, and then **mouse clicking on the Set button**. **Click the OK button** when both points have been set.

Now that the system is set up to record and display the pH data that you want to measure, you need to tell the system whether you want the data to be displayed as in a time versus pH table format or as a plot of pH (y-axis) versus time (x-axis). If you want the data to be displayed in table format, go to the Display section of the left-hand side of the computer screen. **Double left-hand mouse click on the Table line.** A new window will open. Scroll down to pH measurement, and **mouse click OK**. The measured values will be displayed as ordered (x,y) pairs in a time-pH table. Except for the titration curve measurement, which is described in another set of instructions, this will likely be the way that you will want to display all of the measured pH values.

pH TITRATION CURVE MEASUREMENTS

First calibrate the pH electrodes and Drop Counter following the instructions in the pH calibration (first paragraph only) and Drop Counter sections. After the pH electrode and Drop Counter have been calibrated, you will need to set up the system to display the measured data as a pH titration curve, *i.e.*, pH (y-axis) versus volume of titrant (x-axis) plot. Close all of the small display screens on the right-hand side of the computer screen by mouse clicking on the X box in the upper right-hand corner of each display screen. On the far left-hand side of the screen, midway down, you will see a window labeled Display. **Double left-hand mouse click on the Graph line.** A new window should open. **Scroll down to Fluid Volume and mouse click OK**. A graph of Fluid volume (y-axis) versus time (x-axis) should now appear on the right-hand side of the screen. The data that is plotted on the y-axis can be changed **by moving the mouse cursor over the label Fluid Volume** – a box with vertical lines ($\equiv$) should appear. **Left-click on the mouse** – a window with several options should appear. **Scroll down to pH and mouse click.** The y-axis should now be set for pH. The x-axis can be changed in similar fashion **by moving the mouse cursor over the label Time (s)**. A box with several vertical lines ($\equiv$) should appear. **Left-click on the mouse** – a window with several options should appear. **Scroll down to Fluid Volume and mouse click**. The x-axis should now be changed to volume.

PRESSURE MEASUREMENTS

Connect the Pasco Absolute Temperature-Pressure Sensor to the PowerLink unit. The system should automatically recognize what accessories are attached. There really is no calibration per se. You will have to change the units of pressure to atmosphere (the default unit is kPa). To change the displayed unit to atmosphere **mouse click on the Setup button on the top tool bar**. A Setup window should appear indicating that the Absolute Pressure sensor is attached. To the right Absolute Pressure is a shaded box with kPa ▼. **Mouse click on the box**, and then **scroll down to atmosphere**. **Mouse click when the atmosphere line is highlighted**. Close the Setup window and you are ready to start taking measurements.

REMOVING LINES BETWEEN POINTS ON A GRAPH

As tabulated (x,y) data points are sometimes graphically displayed the points are connected by lines. If the data points are not entered in either ascending or descending order the graph looks quite usual as the built-in software connects the points with lines in the order that the points were typed into the table. The lines can be removed by going to the last ▼ button on the graph toolbar, scrolling down to scrolling down to Connected Lines, and mouse clicking to unmark this option.

SIZING AND MANIPULATING GRAPHS

Once an experimental graph has been generated, you have several options for displaying and manipulating the experimental data.

(a) If you want to change the x- or y-axis so as to get the entire graph on the screen, **mouse click on the Display button** on the top toolbar, and **scroll down to Settings option, and mouse click**. The axis settings tab lets you scale the axis to a convenient size.

(b) If you have made multiple temperature runs, and want to select the runs to look at, **mouse click on the Data button on the graph tool bar. Click on what sets of data you want to display graphically.**

(c) If you want to get numerical values from different spots of the graph, **click on the Display button on the top toolbar**, and **scroll down to Measure option, and mouse click.** You will now have a two-dimensional cursor that you can move across the screen to the various points of the curve. Move the cursor to the spot where you want either the x-value and/or y-value, and both values are displayed off to the side.

TEMPERATURE MEASUREMENTS

Open up the Pasco system by mouse clicking on the Data Studio icon that appears on the computer screen. Now connect the Pasco temperature probe (pH/ORP/ISE Temperature Sensor Pas*Port*) to the PowerLink unit. You will want the **Create Experiment** activity. The system should automatically recognize what probe(s) is (are) connected. After clicking on the Create Experiment activity, **Go to the top toolbar and click on the Setup button.** You will want to **click on the Calibrate button to the right of temperature.** Place the tip of the temperature in an ice-water both (which will serve as the 0 °C reference data), **type in 0 for the Point 1 temperature**, wait a few minutes for thermal equilibrium to be reached, and then **mouse click on the Set button.** The lower reference temperature has now been set. For the higher temperature calibration point, use a heated water bath (at water temperature is called for in the experiment). Place the Pasco temperature probe, along with a glass thermometer, into the warm water bath. Measure the temperature of the warm/hot water with the glass thermometer. Now type this value into the box for the Point 2 temperature reading. Wait a few minutes for the temperature probe to achieve thermal equilibrium with the warm water. **Mouse click on the Set button. Now mouse click on OK button.** The Pasco temperature sensor is now calibrated.

Now that the system is set up to record and display the temperature data that you want to measure, you need to tell the system whether you want the data to be displayed as in a time versus temperature table format or as a plot of temperature (y-axis) versus time (x-axis). If you want the data to be displayed in table format, go to the Display section of the left-hand side of the computer screen. **Double left-hand mouse click on the Table line.** A new window will open. Scroll down to the type of measurement (Temperature), and **mouse click OK.** The measured values will be displayed as ordered (x,y) pairs in a time-temperature table.

If you want the measured intensity data to be displayed in a graphical format of time (x-axis) versus temperature (y-axis) double left-hand mouse **click on the Graph button** in the Display section of the left-hand side of the computer screen. Scroll down to the type of measurement (Temperature) you want displayed and **mouse click OK.** The system should now be set up to take make experimental measurements.

APPENDIX B: MICROSOFT EXCEL
COMPUTATIONS

AVERAGE AND STANDARD DEVIATION CALCULATIONS

Some of the calculations during the semester will have to be performed outside of the laboratory time. This will be true of the average and standard deviation calculations based on the experimental data from the entire class. You can perform the calculation using Microsoft Excel spreadsheet. The general computer access laboratories across campus should have Microsoft Excel as part of the standard Microsoft Windows bundle of software. The Excel spreadsheet will appear as a grid of cells, which are identified by both a column alphabetical letter and row numerical value. In cells A5 through A24 enter the mass of the 20 individual post-1982 pennies. In cell A3 type: **=AVERAGE(A5:A24)** . Hit **ENTER** -By typing this command you are programming the computer to calculate the average value of the numbers that are entered in cells A5 through A24. The calculated mean (average value) should now appear in cell A3. To calculate the standard deviation of the numerical values stored in cells A5 through A24, go to cell B3 and type: **=STDEV(A5:A24).** Hit **ENTER** - You have just instructed the computer to calculate the standard deviation of the values that are entered in cells A5 through A24. The standard deviation should now appear in cell B3.

LINEAR LEAST SQUARES REGRESSION ANALYSIS

Microsoft Excel is capable of doing a linear least squares analysis. The Excel spreadsheet will appear as a grid of cells, which are identified by both a column alphabetical letter and row numerical value. Lets assume that you had 5 experimental data points on which you wanted to perform a linear least squares analyses. In cells A5 through A9 enter the numerical x-values of the five data points. In cells B5 through B9 enter the numerical y-values. Once all of the data points are entered, you need to access the built-in statistical package. In cell E1 type- **=INDEX(LINEST(B5:B9,A5:A9),1)** . Hit **ENTER** – by typing this command you have instructed the program to calculate the slope. Note that in the syntax – the y-array of experimental data is entered first, followed by the x-array. The numerical value of the slope should appear in cell E1. To calculate the intercept type in cell E2-**=INDEX(LINEST(B5:B9,A5:A9),2)** . Again hit ENTER. The numerical value of the y-intercept should appear in cell E2. Finally, to get the correlation coefficient, type in cell E3- **=CORREL(B5:B9,A5:A9)** . When you hit ENTER the correlation coefficient should appear in cell E3. The number of data points can be increased. For example, if one had 9 data points, instead of 5, the one-dimensional x- and y-arrays would to A13 and B13, respectively.

APPENDIX C: DENSITY OF SOLID ELEMENTS

Element	Density[a]	Element	Density[a]
Lithium	0.535	Ytterbium	6.977
Sodium	0.971	Neodymium	7.004
Rubidium	1.53	Zinc	7.13
Calcium	1.55	Chromium	7.19
Magnesium	1.738	Indium	7.28
Potassium	0.862	Tin (white)	7.28
Phosphorous (white)	1.828	Manganese	7.30
Beryllium	1.86	Samarium	7.536
Cesium	1.879	Iron	7.86
Sulfur (γ-structure)	1.92	Gadolinium	7.895
Sulfur (β-structure)	1.96	Terbium	8.272
Arsenic (yellow)	2.026	Dysprosium	8.536
Sulfur (α-structure)	2.08	Niobium	8.57
Carbon (graphite)	2.267	Cadmium	8.642
Silicon	2.329	Holmium	8.803
Phosphorous (red)	2.34	Nickel	8.90
Boron	2.46	Cobalt	8.90
Strontium	2.60	Copper	8.92
Phosphorous (black)	2.699	Erbium	9.051
Aluminum	2.698	Polonium	9.20
Scandium	2.992	Thulium	9.332
Carbon (diamond)	3.515	Bismuth	9.80
Barium	3.59	Molybdenum	10.2
Titanium (α-structure)	4.32	Silver	10.50
Yttrium	4.478	Lead	11.34
Titanium (β-structure)	4.507	Palladium	12.023
Arsenic (black)	4.7	Rhodium	12.41
Selenium	4.792	Ruthenium	12.45
Europium	5.259	Hafnium	13.31
Germanium	5.323	Tantalum	16.60
Arsenic (gray)	5.72	Gold	19.3
Gallium	5.903	Tungsten	19.35
Vanadium	6.11	Rhenium	21.04
Lanthanum	6.174	Platinum	21.45
Zirconium	6.52	Iridium	22.55
Antimony	6.684	Osmium	22.61
Cerium	6.77		
Praseodymium	6.782		

[a] Density corresponds to 293.15 K and is in units of grams/cm^3.

APPENDIX D: ALPHABETICAL LISTING THE CHEMICAL ELEMENTS AND THEIR ATOMIC MASSES

Element	Atomic Symbol	Molar Mass	Element	Atomic Symbol	Molar Mass
Actinium	Ac	(227)	Helium	He	4.003
Aluminum	Al	26.982	Holmium	Ho	164.930
Americium	Am	(243)	Hydrogen	H	1.008
Antimony	Sb	121.760	Indium	In	114.818
Argon	Ar	39.948	Iodine	I	126.904
Arsenic	As	74.922	Iridium	Ir	192.217
Astatine	At	(210)	Iron	Fe	55.845
Barium	Ba	137.327	Krypton	Kr	83.798
Berkelium	Bk	(247)	Lanthanum	La	138.906
Beryllium	Be	9.012	Lawrencium	Lr	(62)
Bismuth	Bi	208.980	Lead	Pb	207.2
Bohrium	Bh	(264)	Lithium	Li	6.941
Boron	B	10.811	Lutetium	Lu	174.967
Bromine	Br	79.904	Magnesium	Mg	24.305
Cadmium	Cd	112.411	Manganese	Mn	54.938
Calcium	Ca	40.078	Meitnerium	Mt	(268)
Californium	Cf	(251)	Mendelevium	Md	(258)
Carbon	C	12.011	Mercury	Hg	200.59
Cerium	Ce	140.116	Molybdenum	Mo	95.94
Cesium	Cs	132.905	Neodymium	Nd	144.24
Chlorine	Cl	35.453	Neon	Ne	20.180
Chromium	Cr	51.996	Neptunium	Np	(237)
Cobalt	Co	58.933	Nickel	Ni	58.693
Copper	Cu	63.546	Niobium	Nb	92.906
Curium	Cm	(247)	Nitrogen	N	14.007
Darmstadtium	Ds	(271)	Nobelium	No	(259)
Dubnium	Db	(162)	Osmium	Os	190.23
Dysoprosium	Dy	162.500	Oxygen	O	15.999
Einsteinium	Es	(252)	Palladium	Pd	106.42
Erbium	Er	167.259	Phosphorous	P	30.974
Europium	Eu	151.964	Platinum	Pt	195.078
Fermium	Fm	(257)	Plutonium	Pu	(244)
Fluorine	F	18.998	Polonium	Po	(210)
Francium	Fr	(223)	Potassium	K	39.098
Gandolinium	Gd	157.25	Praseodymium	Pr	140.908
Gallium	Ga	69.723	Promethium	Pm	(145)
Germanium	Ge	72.64	Protactinium	Pa	231.036
Gold	Au	196.967	Radium	Ra	(226)
Hafnium	Hf	178.49	Radon	Rn	(222)
Hassium	Hs	(277)	Rhenium	Re	186.207

Element	Atomic Symbol	Molar Mass		Element	Atomic Symbol	Molar Mass
Rhodium	Rh	102.906				
Rubidium	Rb	85.468				
Ruthenium	Ru	101.07				
Rutherfordium	Rf	(261)				
Samarium	Sm	150.36				
Scandium	Sc	44.956				
Seaborgium	Sg	(266)				
Selenium	Se	78.96				
Silicon	Si	28.055				
Silver	Ag	107.868				
Sodium	Na	22.990				
Strontium	Sr	87.62				
Sulfur	S	32.065				
Tantalum	Ta	180.948				
Technetium	Tc	(98)				
Tellurium	Te	127.603				
Terbium	Tb	158.925				
Thallium	Tl	204.383				
Thorium	Th	232.038				
Thulium	Tm	168.934				
Tin	Sn	118.710				
Titanium	To	47.867				
Tungsten	W	183.84				
Uranium	U	238.029				
Vanadium	V	50.942				
Xenon	Xe	131.293				
Ytterbium	Yb	173.04				
Yttrium	Y	88.906				
Zinc	Zn	65.409				
Zirconium	Zr	91.224				

APPENDIX E – THE SEVEN BASIC CRYSTAL SYSTEMS

The dimensions and volumes of the unit cell (repeat unit) for the seven basic crystal systems are given below:

Crystal System	Unit Cell Volume
Cubic	$V = a^3$
Tetragonal	$V = a^2c$
Hexagonal	$V = a^2c \sin(60°)$
Trigonal	$V = a^2c \sin(60°)$
Orthorhombic	$V = abc$
Monoclinic	$V = abc \sin(\beta)$
Triclinic	$V = abc (1- \cos^2 \alpha - \cos^2 \beta - \cos^2 \gamma) + 2(\cos(\alpha) \cos(\beta) \cos(\gamma))^{\frac{1}{2}}$

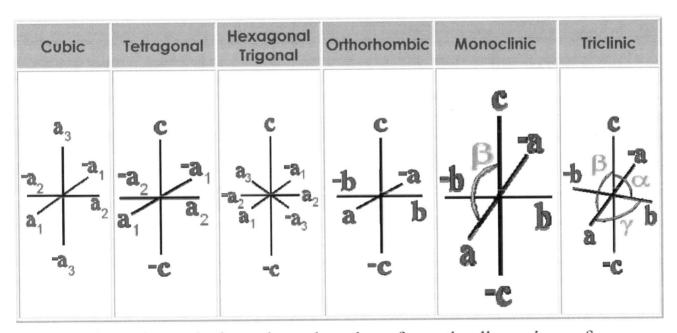

In the volumetric formulas a, b and c refer to the dimensions of the unit cell axes, and the Greek symbols α, β and γ are the inclination angles of the axes in the unit cell.

APPENDIX F: INSTRUMENTATION FOR SPECTROSCOPIC MEASUREMENTS

Most commercial spectrometric instruments have the basic design depicted in Figure E.1. There are numerous variations in the basic design; however, depending on the manufacturer, the wavelength region(s) for which the instrument is designed, the spectral resolution required, and so on. It is beyond the scope of this course to discuss the variations in detail. The presentation here will focus on the basic instrumentation that is typically found in an introductory general chemistry laboratory.

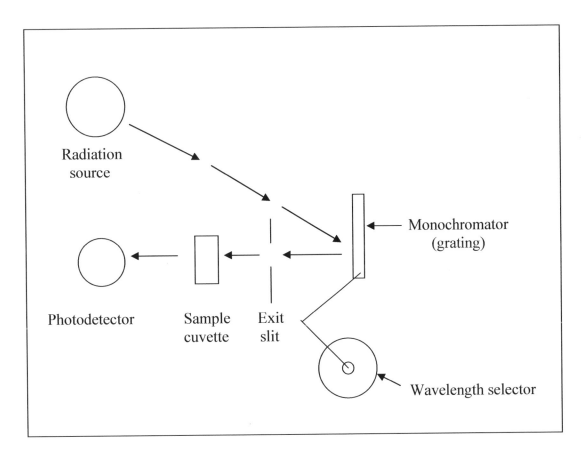

Figure E.1. Block diagram of a typical spectrometer.

Radiation is emitted from the radiation source, which for the visible spectral region is generally a tungsten filament lamp (wavelengths from about 330 to 950 nm). The lamp radiation is directed towards the monochromator (grating), where it is dispersed into the different wavelengths. By turning the wavelength selector, the grating is rotated slightly to change the band of 20 nm of wavelengths passing through the exit slit. The nearly monochromatic light then passes through the sample. Any radiation not absorbed by the sample falls on the detector, where the intensity is converted to an electrical signal that is then amplified and read. For visible and ultraviolet radiation the energy of the photon is sufficient to cause electron emission from the detector's light-sensitive element (e.g., photomultiplier tube) or electron migration in photo-diode arrays, charge-injection devices and charge-coupled devices. The number of electrons ejected from a photoemissive surface, or the charge stored within a collection area, is directly proportional to the radiant power of the beam that strikes the detector surface. Infrared radiation is quantified using a thermal detector. Here, the measured property is temperature dependent, such as the resistance of a strip of nickel or platinum metal (thermister). These metals exhibit a relatively large change in resistance as a function of temperature.

The Pasco colorimeter used in Experiments 19, 20, 21 and 26 functions in a similar manner, except that the radiation source consists of four light-emitting diodes that emit red, green, blue and orange light. The spectral resolution of the Pasco colorimeter is considerably less than the more expensive spectrometers used in the upper level analytical chemistry and instrumental analysis courses. For our purposes, spectral resolution is really not much of a concern because great accuracy is not needed to demonstrate the basic chemical principles associated with each laboratory experiment. For the experiments that are being performed in general chemistry, good results can be obtained with the Pasco system.